Environmental Management Systems

A Step-by-Step Guide
to Implementation and Maintenance

Third Edition

Christopher Sheldon and Mark Yoxon

publishing for a sustainable future

London • Sterling, VA

Third edition first published by Earthscan in the UK and USA in 2006
Reprinted 2008

Second edition 2002

First edition 1999

First and second editions were published with the title:
Installing Environmental Management Systems: A Step-by-Step Guide

ISBN 978-1-84407-257-6

The authors have made every effort to provide accurate and complete information. No
responsibility, legal or otherwise, is accepted for omissions or errors. The publication is designed
to provide a practical handbook for those implementing an environmental management system.
It is not intended to render legal, accounting or any other professional services.

Typesetting by JS Typesetting Ltd, Porthcawl, Mid Glamorgan
Printed and bound in the UK by TJ International, Padstow
Cover design by Andrew Corbett

For a full list of publications please contact:

Earthscan
Dunstan House
14a St Cross Street
London, EC1N 8XA, UK
Tel: +44 (0)20 7841 1930
Fax: +44 (0)20 7242 1474
Email: earthinfo@earthscan.co.uk
Web: **www.earthscan.co.uk**

22883 Quicksilver Drive, Sterling, VA 20166-2012, USA

Earthscan publishes in association with the International Institute for Environment and
Development

A catalogue record for this book is available from the British Library

The paper used for this book is FSC-certified.
FSC (the Forest Stewardship Council) is an
international network to promote responsible
management of the world's forests.

Mixed Sources
Product group from well-managed
forests and other controlled sources
www.fsc.org Cert no. SGS-COC-2482
© 1996 Forest Stewardship Council

From Mark
For Mo

From Christopher
For my mother, my father and, as always, Jennifer

Contents

DOING

Figures, Tables and Boxes

Boxes

Acronyms and Abbreviations

BAT	best available technology
BPEO	best practicable environmental option
CBA	Cost benefit analysis
CEN	Comité Européen de Normalisation
CFC	chlorofluorocarbon
CMA	Chemical Manufacturers Association
CSR	corporate social responsibility
EAC	European Accreditation of Certification
ECI	environmental condition indicator
EMAS	Eco Management and Audit Scheme
EMP	environmental management programme
EMR	environmental management representative
EMS	environmental management system
EPE	environmental performance evaluation
EPI	environmental performance indicator
EVABAT	economically viable application of best available technology
HR	human resources
IAC	International Accreditation Forum
IER	initial environmental review
LCA	life cycle assessment
MPI	management performance indicator
OPI	operational performance indicator
PDCA	Plan-Do-Check-Act
QMS	Quality Management System(s)
RAL	requisitory assessment level
REPA	resource and environmental *process analysis*
SME	Small and Medium Sized Enterprises
TC	Technical Committee
TNA	training needs analysis
TQM	Total Quality Management
VOC	volatile organic compound

1

Introduction: How to Use This Book

Is this book for me?

Unless you're the sort of person who thinks that the Introduction in a book is a waste of paper, then you're probably dipping into these pages wondering if this is going to be a useful book or not. The title probably gave you a hefty clue, so it's more than likely that you are going to be interested in environmental management systems (EMS), their installation, implementation and maintenance. It's possible that you've actually got the job of installing one, running, updating or perhaps training people who have to do the job themselves. Whatever your reason for picking up this handbook, we guarantee you're going to find something useful and/or usable in it.

Over the last 15 years, we've had the privilege of observing a lot of managers from a wide variety of functions grappling with environmental management while they attended training courses that we were running on the subject. The first day always seems to be the hardest; a lot of new information to take in, some of it politely described to us as 'very dry indeed'. Then on the second day, after a few exercises, delegates begin to act with more confidence around the subject, mainly because they've seen just how far their own wisdom can actually take them. Invariably, by the third day we can hardly hear ourselves think for the sound of open palms smiting foreheads (delegates smiting their own foreheads you understand, not each other's) as they 'get it'. In training circles, it's known as the 'Aha! factor'; that blissful moment when all the fragments of information fall into a pattern and the subject begins to make sense. This book tries to capture some of those 'Aha!' moments, and pass them on to those that need them.

In this opening chapter, to clear the ground, we're going to answer some of the basic questions about EMS, their implementation and running, and lay out the aims of this book and the way we're going to approach the subject. This will let you check the level of your understanding, maybe challenge a few of your assumptions, and reveal some of ours. Mainly though, it'll ensure that you haven't bought this book thinking that it's something to do with garden design. If you haven't slapped your forehead once by Chapter 4, then you probably skipped this introduction.

What this introduction covers
- What is environmental management?
- Why bother with it?

- How does it work with environmental regulation?
- What is an environmental management system?
- How will your culture affect your systems?
- Why is sustainability development important to my organization?
- How to get the best from this book

What is environmental management?

If financial management is managing the finances of a company, and quality management is managing the quality of its products and processes, then it stands to reason that environmental management is managing the environment that the company operates in. Well ... yes and no.

It's been a steep learning curve, but most of us now accept that everything we do affects the environment. It even extends down to the way our bodies are dealt with after death. To illustrate the point, a local authority discovered during a review of its environmental effects that its most significant impact on the environment was the air emission from the crematorium chimney. It's reputed that the report went on to suggest a return to landfill!

Obviously, some effects are bigger than others (one person breathing, say, is less of an impact than making steel), yet each of these effects is interrelated. That relationship can sometimes disperse the impact and lessen it, while in other cases it can multiply the strength of the impact in a cumulative manner. In one sense, it is true that the whole interaction of individuals and the environment is so complicated that we have to accept that we cannot control or 'manage' the effects. We may never be in a position to manage something as complex as a planetary ecosystem without unforeseen results and/or problems arising. Managing the environment is thus a contradiction in terms, even without the added complexity of ideas like managing for sustainability.

Accepting this, environmental management does not seek to manage the environment directly. Instead, it concentrates on the more indirect, but nonetheless effective, route of managing an organization's activities that give rise to impacts upon the environment. This is more than playing games with words just for the fun of it. The semantics are important. The focus of the work becomes the interaction between the organization and the environment, and the rather fluid interface between the two. It is the environmental aspects (as opposed to the financial or quality aspects) of an organization's activities, products and services that are the focus of management.

Why bother with it?

All biological organisms rely for their survival on the effectiveness of their feedback loops to give them clear and accurate information on the world around them. Evolution displays a tendency over time to favour those organisms with better feedback loops. It appears that the more one can sense, the better the chance of survival. Corporate organisms display the same characteristics as their biological relations.

A management system can be seen as a way of improving (or even establishing) these feedback loops in an organization. An EMS specifically improves the feedback about a constantly evolving area: environmental protection. Continuing social awareness concerning the state of our environment is another aspect of 'sensitization' caused by better feedback loops. Obviously the more finely attuned an organization is to new developments, the better placed it is to react, to plan and to improve ahead of any legal or market requirements. An additional benefit to corporate organisms is that the quality of the information received and acted upon is directly related to the efficient use of resources in meeting the new demands. Priorities vary according to the circumstances of the organization and its relationship with its public.

An EMS can help to define circumstances of which others may not even be aware, keeping the more sensitive organization continually at the head of the corporate food chain. How much do you need to know about environmental issues? More than your competitors; and you only get better knowledge from better feedback loops.

How does it work with environmental regulation?

One argument which is heard quite frequently is that environmental management is not necessary if one is already following all the relevant environmental regulations that apply to the site or the operations in question. This is specious, much akin to saying financial management is not necessary if you're paying all the right taxes. In order to make sense of how environmental issues have changed our understanding of business management, one first has to appreciate where regulation leaves off and self determination (in terms of voluntary self-regulation) starts.

Two types of mechanisms have evolved in society in order to express approval or disapproval. One is the legislature and associated regulations; the other is a variety of instruments loosely grouped under the heading of 'the market'. In societies where 'command and control' of environmental impacts is the leading principle, the law finds expression through a dense framework of regulations. These not only determine specific baseline environmental performance parameters but may also seek to be quite prescriptive as to how that performance will be achieved. The more regulations there are, the less room there is for organizational management to use alternative methods to achieve the same ends. It also means that 'environmental' acceptability is being closely defined by the representatives of democracy, rather than by the individuals of that democracy.

Using the market as a feedback mechanism is favoured by some national policy makers as a more fluid response to such an uncertain area as environmental issues, allowing organizations more flexibility in the way they meet agreed targets. As a result, it is incumbent upon individual organizations to take account directly of the expressed preferences of their own market, while still using the law as a baseline for performance standards.

Self-regulation (industry agreeing to work to methods and standards beyond legal requirements) can be seen as one attempt to fill the credibility gap between the market and the legislative framework; the gap between what the people's representatives have a mandate to ensure, and what the people themselves want. The expression of an individual person is a markedly more complex set of signals to read and as a result is more open to interpretation. Thus, individuals who have differing expectations of what they require from an organization can be identified as being members of different groups known collectively as 'stakeholders' (see Chapter 2). It is taking these widely divergent groups into account, and attempting to meet their expectations on a continuing basis that makes the establishment of a formal management system an increasingly useful tool for all levels of management.

In the 1980s, the market started to express a preference for environmental awareness as a corporate characteristic. No one would deny that the amount of environmental legislation has increased considerably since this period, though definitive figures are harder to agree upon. Most observers agree too that this marked preference for a regulatory response to environmental problems is an expression of the market's unwillingness to believe that industry could possibly put its own house in order. There are not many organizations, however, that are content to be driven by the threat of legislation, especially if they wish to be around to reap the benefits of a relatively short term 20 year business plan.

Environmental issues are complex in and of themselves, let alone the exponential complexity caused by their interaction. Given that these issues are also currently being legislated for in terms of strict liability, it is hardly any wonder that industry has developed management tools and technical standards that will help it allow managers to manage. The aim is to manage proactively, to take action in advance of legislation, exceeding social expectations before they are realized in the form of a regulatory 'blunt instrument'.

What is an EMS?

A key component of industrial self-regulation is the development of technical standards at national, European and international levels. In recent years, these standards have moved from mutually recognized technical details concerning the construction and performance of specific products or components, through to the development of standards on management systems, and the interaction of their individual elements.

There is no mystery about 'management systems'. Even a sole trader will have some form of management system: it may not be formal, or based on standardized formatted paperwork, but it will be a management system nonetheless. In short, businesses only survive because they have some sort of system that works. No two management systems are the same, because no two companies are the same. Yet, at heart, any management system is simply a way of moving information around inside an organization. Its job is to make sure that the right information arrives at the right place at the right time, so that the right decisions can be made. On the other hand, in order to achieve this seemingly simple end, it will need to take into account an organization's personnel, structure, planning functions, operations, processes, procedures and even its habitual practices.

When the concept of producing technical standards that would define management systems first arose in the 1970s, the aim was to publish a document that would record all the landmark activities and functions that made for a successful system. The system would in turn deliver the outcomes that its management had identified as a desirable series of objectives. At the time of publication in 1979, British Standard BS 5750 was the world's first national standard on quality management systems, or indeed, any type of management system. It provided a remarkably long-lasting model, and is still in use today as the international standard ISO 9000.

Many will already have heard of ISO 9000 and even more will have had experience of working within a management system designed according to its principles. Such experience will prove useful when looking at environmental management systems. What will be of most interest to readers of this book will be the environmental equivalent of ISO 9000, called *ISO 14001: Environmental management systems – specification with guidance for use*. In Europe, the European Commission decided to encourage the take-up of a similar, but in some ways more prescriptive, voluntary scheme by promoting something called the Eco Management and Audit Scheme (EMAS). ISO 14001 and EMAS have done a lot to provide a focus of activity in the development of EMS, and it is likely that many of the readers of this book will be seeking to install a system that ultimately conforms to the requirements of one or both of the schemes. Even national or regional 'phased implementation schemes' for EMSs (such as those relating to British Standard BS 8555), which were originally designed for smaller enterprises, are ultimately aligned with ISO 14001. There is a commentary on the requirements of all the relevant schemes in Appendix I. Throughout the main text of the book, we have referred to the generic term 'EMS', encompassing all the schemes, only making specific references to the schemes for clarification where necessary.

This book has been designed to help those of you installing and maintaining all or even part of a management system whether or not you are planning to use one of these schemes. If you are, it should be remembered that the context for their development was to enable organizations to install and develop systems appropriate to their own organization and the way it works. In other words, implementation of an EMS should focus on the whole idea of putting the 'self' in environmental self-regulation.

How will your culture affect your systems?

Formal management systems tend to be regarded as a subject second only to watching paint dry in most organizations. Working with management systems lacks the collective romance of the struggle for corporate survival, dispenses with the thrill of managerial fire-fighting and singularly fails to feed the need in every individual to make a useful (and noticeable) contribution to organizational profits. Yet, with the latest generation of formal systems to manage the environmental impacts of organizational activities, products and services, there are now managers who can bring to the subject the experiences and models offered by the earlier versions as well as the introduction of quality management systems. This practical help will tend to overcome the inertia that can sometimes anchor such work to the drawing board. Such managers are also well aware that it is possible for the organization to take over the system, as well as for the system to take over the organization.

In earlier models, any initial reservations concerning the potential downside impact on the company were countered by claims that a positive management 'culture' would be created in the simple act of management system installation. As evidence that this does not happen automatically, one need only look at how the philosophy of 'Total Quality Management' (TQM) has somehow foundered in many companies.

Corporate historians are already recording that where TQM failed, organizations had mistaken the means for the goal, and the management systems that gave shop floor reality to TQM were mistaken for the end result. What few had apparently taken into account was the pre-existence of a culture within an organization that would have a basic influence on any formal management system; namely that the system would give voice to the culture, not the other way around.

As a result, we now recognize that the cultures of individual organizations, sites or even departments can affect how a system is used. Where systems are being used to little or no effect, many point the finger at the system itself certain that there is a basic design flaw in some or all of its parts. Those who have already suffered at the hands of rigidly designed and bureaucratically executed systems will variously cite senior management, the writers of ISO 9000, national culture or some other intangible reason beyond their responsibility and outside their control.

In reality, those who have had the opportunity to establish or assess management systems over the years have a remarkable congruity of observations. With deference to all of those who knowingly or otherwise contributed to this book, our distillation of their comments can be laid out in the following models:

Name: Dictatorship
Belief: The individual will make mistakes if left to their own devices
System profile: Command and control, heavily bureaucratic, extremely formal

Name: Natural Selection
Belief: Design engineers will make mistakes which are then replicated throughout company processes
System profile: Product and process focused, inflexible on a day-to-day basis but changes with each new product

Name: Survivalist
Belief: Mistakes are allowable as long as a profit is shown at the end of the financial year
System profile: Essentially market driven, operated by maverick managers who use formal systems only as a post hoc rationalization of their own actions

Name: Learning
Belief: Mistakes provide important feedback. People need structures to direct, not constrict, their abilities
System profile: Flexible, constantly under revision to reflect both the needs of best practice and those of managerial policies within the organization

You may recognize your company in one of the above classifications, or perhaps your own department, or maybe yourself. You may even have experienced two or more in the same organization. The dominant culture will be created from the top down and over a period of years, so for those attempting to create a management system for environmental issues, it is worth considering how your own culture may shape the system.

All this is both good news and bad news for a manager looking to implement an EMS. The good news is that those in the organization opposed to a change in culture need not worry. The entire workforce undergoing an environmental 'Road to Damascus' experience is not an inevitable consequence of contact with an EMS. The bad news is that EMS implementation project managers have to be aware of what the existing culture will do to subvert a new management system. There is a need for fine judgement during implementation as there are times when adaptation of the system may be valid to ensure that it is efficient. The key question to ask under such circumstances is: 'Is it effective?' rather than: 'Is it convenient?'

In the long run, any organization that wants to get the most from a formal EMS is going to have to be prepared to engage with the complex issues that require management. Decisions will not necessarily become 'easier' (indeed in many cases they may actually become more challenging), but they will be decisions that are made in the light of full knowledge of the implications of agreed actions. In short, the outputs of an EMS will reflect the values of the organization; they will not gain any additional 'ethical pump-up' from the essentially value-neutral EMS. Over an extended period of time, an EMS may have an effect on the culture, but this will be an evolutionary and controlled process.

Phased implementation schemes can also be useful in controlling the pace of change inside an organization. It means that within the confines of the market driven approach, the pace of system installation can be matched to what the organizational culture is prepared to accept. If project managers suspect that there may be some high levels of resistance to change, phased approaches (such as those typified by the UK's Acorn/BS 8555 scheme) may be worth a closer look. See Appendix I for further details.

Why is sustainability important to my organization?

The local, regional and global environments are only some of the many aspects that an organization alters by carrying out its activities. Environmental protection has also been subsumed into a larger concept known as 'sustainable development', and many organizations find that in tackling environmental issues, questions are raised about the whole way their company interacts with other facets of society and the world around it. Einstein is reputed to have defined the environment as 'Everything that isn't me' and while accurate as far as it goes, the definition glosses over the fact that each of those elements is changed by interaction with the other. For an organization attempting to thrive and prosper, it makes sense to sustain the environment that is supporting it; but when the definition of the environment is 'everything that isn't itself', it begins to include factors traditionally left to other social institutions; hence the growing interest in the broader concept of sustainable development.

Like most concepts that are popularized by international commissions, sustainable development is expressed in a form of words to which everyone can subscribe. The Brundtland Commission was set up by the United Nations in 1987, and its final report took the sustainable development concept which originated in the 1970s and used it as a bridge between those who were polarized about whether economic growth was essentially good or bad. It gave expression to the idea as:

> . . . development which meets the needs of the present without com-
> promising the ability of future generations to meet their own needs.

Such a definition draws upon the ability of diplomatic language to mean all things to all individuals, which in turn has led to the wide adoption of the idea around the globe and several hundred definitions of sustainable development. Despite the plethora of subsequent 'clarifying' definitions, it is generally accepted to include basic principles of environmental protection, quality of life, social equity and futurity. Its flexibility is its great strength.

Unfortunately, in our greatest strengths lie our greatest weaknesses. While sustainable development is a phrase that encourages little dissent from such apparently widely diverging sectors as economists and environmentalists, it is also dogged by the fact that everyone is agreeing to their own version of that concept. In addition, the business contribution to sustainable development has been further defined as a quality of continuance, otherwise known as sustainability. Definitions are haggled over, experimentation is rife, and many commercial and industrial organizations are hard at work attempting to bring the concept into the world of everyday business decisions. Common guidelines for management on the subject are beginning to emerge but they are far from being a 'magic bullet'. Some organizations are having more success with the idea than others and, for many, environmental protection is the first rung on the ladder.

It is easy to see why. For many, the inescapably fundamental nature of the environment and our reliance upon it makes it our greatest priority. Social networks and institutions may break down and reform; not without cost, but at least new models and techniques emerge. Economic systems may strengthen or weaken their global bonds and new models offer more flexible approaches, but even at the level of barter, something will survive even in the most extreme cases of economic breakdown. If our environmental systems fail to support us, however, the human species' ability to adapt to sudden and extreme changes may not be sufficient to guarantee our survival. Organizations and individuals find such thinking enables remarkably straightforward prioritization.

Even so, there are still business managers who see the environment as something of an externality that has drawn fashionable attention to itself in a welter of uncertain science and collective angst. As such, to them, managing for sustainability (whether social, economic or environmental) will be little more than an unwelcome additional responsibility among their more important mainstream management activities. On the other hand, those managers looking forward will be able to see the possibilities of an entirely new way of making business decisions, bringing together several strands of mental and physical activity previously excluded from the strategic priorities and solution seeking. The fact that business is being asked to take on these problems can be seen as a managerial burden and an unnecessary addition, or it can be seen as further proof that organizations in the industrial and commercial sectors provide useful models for successful problem solving and solution delivery. Far from a penalty, being asked to take account of new factors is a reward; a sign of success in dealing with a complex system of economic factors so well in the past. In essence, the 'responsibility' of business is increasing, because its 'ability to respond' is also increasing.

This means that organizations who are benchmarking themselves against their competitors (or even their collaborators) are seeing that the sphere of activity is expanding outwards.

The new factors of equity, futurity and ecology will have to be embraced or the organization will run the risk of falling victim to competitive extinction at the hands of those who can manage them successfully. An EMS is a way, not only of recognizing new and uncertain elements that require management, but of constantly remaining in a state of readiness for the better class of problems that emerge with each successful decision.

How to get the best from this book

This book has been designed to be a comprehensive guide to implementing an EMS. It can be used by anyone: from managers of large manufacturing enterprises through to one-person service operations and any variant in between.

It explains how to:

- review what you need from an EMS;
- design and run an EMS to your own specification;
- incorporate the requirements of international and European standards (ISO 14001/EMAS);
- plan, manage and deliver your own EMS Implementation Project;
- avoid unnecessary paperwork;
- make sure your organization benefits directly from an EMS;
- get early and beneficial results with focused effort.

The book has been written with the busy manager in mind, and will take you step by step through the process of designing, installing and maintaining an EMS. The main body of the book is given over to 14 chapters that cover the basic principles of EMS implementation and management. Each chapter includes helpful guides with hints, tips and practical 'insider' information on the whole approach to managing your organization's environmental aspects, with a joint emphasis on both fun and profit.

Each chapter contains:

✓ ISO/EMAS quick check

For those implementing an EMS with a view to meeting standardized requirements, this is an at-a-glance guide to the parts of ISO 14001: 2004 (the international standard for environmental management systems) and EMAS: 2001 (the European Union's Eco Management and Audit Scheme) that are either referenced directly or will benefit from the tasks in the chapter.

◻ Chapter executive summary

A quick guide to the subject matter and the task themes in the chapter.

⮩ You are here

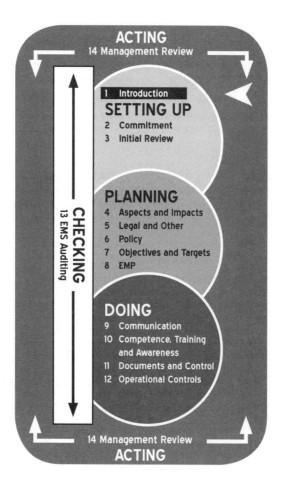

A diagrammatic representation of where a specific chapter stands in relation to the implementation process. When you're looking at the detail of an EMS, it's very easy to lose track of where you are and how you stand in relationship to the other elements of the system. This quick visual reference will help to keep you on track, and provide you with an at-a-glance guide to where you are, and how far you've come.

Toolkit requirements

An assessment of the skills and competencies required to complete the work of the chapter. These skills may already exist within your current workforce or implementation team. If they do not, your choice is to grow some kind of in-house capability or buy in the skill via a consultancy.

Context

The subject matter of the chapter put into the context of the implementation project/process.

✍ Tasks

A series of explanatory paragraphs interspersed with tasks that will need to be completed and carried out in order for the implementation of the EMS to take place. All the tasks are appropriate, though the time it takes to complete the task will often depend on the number and nature of environmental impacts that need to be managed. The text has been prepared in such a way that it may be read through once in narrative format, and then used as reference material directed by the pertinent 'signpost' questions.

💭 Narratives

Examples of what the processes in the chapter can look like in real life, drawn from actual implementation experiences in real organizations. Here, we will start the ball rolling by including brief profiles of the two fictional companies we have created to illustrate common problems. Each one has been created in order to demonstrate particular characteristics and problems (resources available to organization, manufacturing or service focus, employee base and so on). They are in no way to be taken as exemplars. We recommend that, whatever your situation, you read both sets of narratives, simply because your individual circumstances may not debar you from the problems suffered by either company. Any similarity between these companies and any organization (living or dead) is entirely coincidental:

Smallco - Profile

Smallco is a small, well-established, family owned company. They occupy two units on an out of town industrial estate and employ 67 staff. Their core business is the design and production of metal fabrications using mainly aluminium and steel. Smallco buy in a variety of raw materials in the form of finished aluminium extrusions to mild steel stock in various forms from sheet to angle iron and strip metal. Stock levels of all metals are high and not particularly linked to booked work. Their customer base includes the automotive industry, the electronics industry, the commercial building trade and a range of small customers with individual and diverse requirements. Since 1999 they have operated from a medium sized, partly modified site on a 1960s industrial estate. The site, a former biscuit factory, is not really designed for the work Smallco are doing, but the local council – keen to encourage new jobs in the town – were instrumental in persuading Smallco to locate to the town. The MD is a local magistrate and they sponsor the local estate football team, most of whom work for the company.

Over the last three years business has grown significantly. Smallco have coped with the demands for increased orders by instigating a 16 hour working programme on two shift patterns. Employment has grown by over 40 per cent to cope. The site is now very cramped with poor storage facilities. Smallco management have outline plans to relocate to a larger, purpose built site at a new industry park and have begun exploratory talks with the local council to investigate their options. The short contract nature of much of their business means finding the investment capital is not easy. Further expansion of business is planned for the next fiscal year by further extending their customer base. Smallco are under increasing external pressure to adopt an EMS. Two major suppliers, both recently certified to ISO 14001, have indicated that they now expect suppliers to satisfy a ten point environmental management checklist derived from one of the larger trade associations. One of them has indicated their willingness to lend Smallco support to a phased EMS implementation scheme. Smallco have a Quality Management System (QMS) in place, but are a little confused by these new environmental demands and have poor understanding of just what this means. The quality manager has been briefed to investigate the demands from suppliers and just what EMS means in operational terms.

Early findings indicate the following:

- The customers are very serious, committed and wish to protect their own certification as apparently one of the clauses of ISO 14001 suggests suppliers must be considered. The issue is unlikely to go away.
- Several letters are on file from the local environmental health officer which indicate local political support and encouragement for environmental management.
- The Operations Manager at Smallco has previous experience of EMS from their last job.
- There appear to be business benefits in the form of cost savings from adopting a proactive approach to environmental management.
- There may be opportunities to extend the QMS to embrace environmental management.

The quality manager will report back at next month's management meeting. Alongside the points above, the key recommendation will be that for the first phase of work, the QMS can be mapped against ISO 14001 and this model used to establish a simple system for Smallco's first approach to EMS. He will also investigate the phased implementation scheme, which can also be used to take the company towards ISO 14001, in more detail. Using the international standard now will mean not having to undo a system later if certification becomes an option or necessity.

BIG Inc - Profile

BIG (Business Information Guidance) Inc is a large and commercially successful information technology consultancy which supplies both hardware and software solutions from a range of market options for clients all over Europe. They design and install networks, help form IT strategies, design intranets, web sites and provide associated custom software. They are a relatively young company having been formed in the late 1980s by a core team of technicians and marketing students, who are now the senior management.

They have 400 full time employees, but another 600 are contracted 'associates' who work in consultancy and software design. BIG Inc leases a medium sized office block in an urban redevelopment area in a provincial city, housing 350 personnel, with the remaining numbers either telecommuting or 'hot desking' on site. The company likes to use the latest technology and techniques including remote access meetings, virtual offices and 'intelligent' electronic mailing systems. There are also five more leased office suites in key centres around Europe.

The founding managers of BIG Inc are particularly proud of their business culture, which they feel is more direct, more personally involving than the average company, reflecting their original values. Their motto is 'Just Do It – Better!' Consequently, in response to some interest from customers, the need to differentiate their products and forwarding their stated values, they have appointed an 'associate' as a project manager to achieve ISO 14001 registration within 12 months, and to consider EMAS after that.

Trade secrets

A series of hints and tips that will make life easier, based on our implementation and training experience.

Things to think about

Some pointers and thoughts as to how business and environmental issues might evolve. These can be viewed as opportunities or challenges, depending on your current state of mind.

There is also a series of Appendices containing useful background information:

Understanding the requirements of EMS

A commentary on the basic requirements of ISO 14001 and EMAS, including the structure of phased implementation schemes (such as those based on British Standard BS 8555).

Third party auditing

An outline of the process used by independent bodies in assessing EMS for certification to ISO 14001 and/or verification of an EMAS public statement. Inspection regimes for phased implementation schemes are also discussed.

EMS project management

In larger companies, it is likely that the implementation of the EMS will take the form of a project and will probably require the assembly of a special implementation team. This appendix contains valuable project management information, specifically geared to EMS installation.

Planning for continual improvement

Both ISO 14001 and EMAS use 'continual improvement' as a guiding principle. This appendix contains vital information on how such a principle will need to be incorporated in strategic management functions.

Integrating management systems

Management systems already exist for defined aspects of operations such as health and safety or quality as well as environmental impacts. Increasingly, organizations are developing parallel systems to run processes managing Corporate Social Responsibility and sustainability issues. Whatever other management systems your organization runs, this appendix discusses some of the specific issues that apply to the integration of all systems not only to avoid duplication and oversights, but also to maximize the potential efficiencies and strategic business benefits.

Glossary

In any new discipline, there are new technical terms, acronyms and mnemonics that will not be readily identifiable. The glossary contains the most important terms you are likely to need.

Authors' note

We have kept the emphasis throughout on the practical and the pragmatic, providing as much information as we thought would prove useful, but without overloading the prospective EMS implementors. You will get the best from the book if you undertake the exercises and tasks that have been laid out for you in the sequence in which they are presented. This will ensure that you get the cumulative benefit of what you discover about your own organization and circumstances. Some schemes that include external recognition of the implementation work on a stage-by-stage basis may have more detailed requirements in terms of producing objective evidence that can be verified by scheme inspectors. This book can still be used to drive the implementation process, as the same basic principles apply to all EMSs.

The guidance in this book is perforce of a broad nature. We are passing on what we know has worked for other individuals and companies, but following such guidance does not absolve you from meeting your own and your organization's environmental liabilities, which arise from your particular activities, products and services. You will need to decide what works for you from the information in this handbook, using it as a starting point, not a final destination. We are keen to see environmental management systems up and running, but not to the point of making ourselves 'human shields' in any potential law suits.

In the spirit of continual improvement, we are always pleased to hear from those of you who are using the book, especially where you think that you have a tip that would help other people. If you do have any further suggestions or ideas for inclusions in later printings of the book, please write to the authors c/o Earthscan Publications Ltd or e-mail your contribution to greeninck@btconnect.com or markyoxon@inform-global.com. Unfortunately, we cannot undertake to answer all your questions, but we'll do our best.

Setting up

2

Commitment: Do We Want to Do This?

An EMS is a formal approach to managing the aspects of an organization's activities, products and services that have, or could have an impact on the environment. It can be used to prioritize actions and resources, increase efficiency, minimize costs and lead to better, more informed decision making. One of the most important differences between a successful EMS and one that ends in failure is commitment to making it work in the first place. This commitment has to come down from the top of the management structure and, to do that, the strategic reasons for having an EMS have to be clearly understood, even if only limited resources can be spared for a phased implementation approach.

✓ ISO/EMAS quick check

Area of EMS	ISO 14001	EMAS
Environmental policy	4.2	Annex I-A.2
Resources, roles, responsibility and authority	4.4.1 (Para 1)	Annex I-A.4.1
Management review	4.6	Annex I-A.6

☐ Chapter executive summary

Move to the next chapter when you can answer all of the following questions in ways which make sense to you and your organization:

Have you characterized your top level commitment?

Do you appreciate the role and influence of stakeholders?

Do you understand how to define the issues?

How does your supply chain affect an EMS?

Is external recognition appropriate for your EMS?

⌕ **You are here**

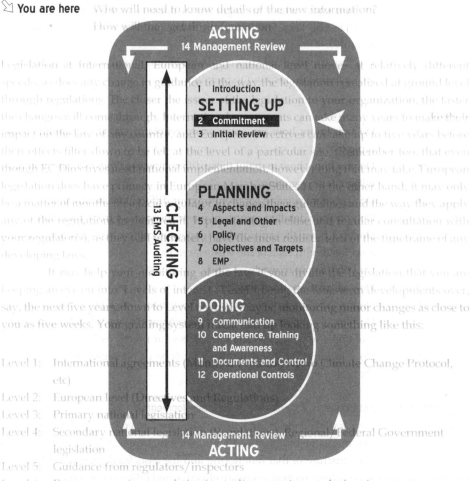

Level 1:   International agreements (... Climate Change Protocol, etc)

Level 2:   European level (Directives and Regulations)

Level 3:   Primary national legislation

Level 4:   Secondary national legislation ... Regional/Federal Government legislation

Level 5:   Guidance from regulators/inspectors

Level 6:   Rumours, gossip, speculative journalism, passive smoke breaks

⚒ Toolkit requirements

Knowledge

- your organization's structure, including top management roles and responsibilities;
- your parent organization's structure, including top management roles and responsibilities (if applicable);
- your organization's business plan and priorities;
- benefits of improved environmental performance for your organization;
- consequences of failing to improve environmental performance;
- relevant legislative development;
- relevant environmental management schemes, guidance, best practice and standards.

Skills
- analysis and data evaluation skills;
- communication and presentation skills (written/verbal).

💡 Context

In terms of an EMS, the word 'commitment' is usually found in the same sentence as the phrase 'top management'. It is frequently used to indicate the relative importance of strategic issues in environmental management and to emphasize that no EMS installation can be completely successful without this top level involvement. The reasons are not hard to divine. A new management system can mean a fundamental change in the way that an organization works both physically and mentally. Senior management have a vital role to play at any time of change within their organization, but particularly so when attempting to install a new formal system of managing the environmental aspects of their business which will affect, to a greater or lesser extent, every function.

It is not just during the installation process, however, that senior managers influence how an EMS is used. The continuing health of the system, once it has been embedded, will be in their hands through the mechanism of the management review (see Chapter 14). Even on a day-to-day basis, their decisions will have a direct impact on the management system itself, and will either support it or circumvent it. For many, the change of approach may be gradual; evolutionary rather than revolutionary, especially if the implementation route makes use of a phased approach (such as that outlined in British Standard BS 8555). The important factor in terms of EMS implementation is being able to identify the nature and level of commitment available.

Up until relatively recently, business factors that have required decisions have lacked volatility; the last 60 years or so have been remarkably stable in that regard, with the focus on technology for the most part. Successful manipulation of the factors controlling finance, human resources (HR), raw materials, marketing and quality controls have been the basic ingredients of a thriving company. Since the advent of environmental management in the mid 1990s, a generation of executives is having to incorporate these new requirements into their everyday existence. For some this means having to change the way they think about certain problems in the first place, let alone attempting to balance a whole new set of priorities. These priorities require a certain amount of flexibility; a state that is not always achieved overnight.

As a result, it is important for anyone involved in EMS implementation to gain the solid commitment of the organization's top management, know the nature and level of that commitment and understand how it will continuously affect the implementation and maintenance of an EMS.

✍ Tasks

Have you characterized your top level commitment?

Over the last decade, there has been an increasing number of business surveys, probing attitudes towards environmental management. These have taken place at local, regional, national and international level, within specific industry sectors, and across sectors by size of workforce or financial turnover. Taking an overview of these surveys, managing directors, owner/managers and sole operators alike have been remarkably consistent in the reasons they have cited for addressing environmental management. The most popular motives for taking up a formal EMS include the following:

We want to avoid prosecution. It would be hard to be a manager of a modern organization and not be aware of the increase in environmental legislation (see Chapter 5). Yet any management team wanting to avoid both financial penalties and damaging publicity that arises from prosecution are focused on only one side of the environmental equation. In focusing on the law, they will probably be oriented towards legal compliance and show little interest in taking their environmental performance any further than that. In such circumstances, trying to get, say, an energy management initiative off the ground is going to be difficult. While it may save money, it is not a legal requirement.

We need it because our customers are asking us to get one. Management responding to the requests of clients and customers can be the best and worst kind of committed individuals. At best, they are acting ahead of any specific demands, but can see that the marketplace is moving gradually in the direction of seeking reassurance in the way the organization tackles environmental issues. At worst, they will be dragging their feet, seeing little motivation to enter into the spirit of the initiative, having felt that a large client is holding a gun to their head and forcing them to get a formally recognized EMS. Such grudging acceptance will make managers immune to arguments other than the fact that an EMS has just become the latest 'licence to trade'. (See also 'How does the supply chain affect your EMS?' on page 25) On the other hand, such management may be more open to committing to a phased implementation scheme, where they may feel that, at most, they are only committing themselves to reaching an appropriate phase/stage.

We want to save money. Seeing an EMS as an excuse to make greater efficiencies and thus save money, while appealing to a broad cross-section of top management, will skew the development of objectives and targets in favour of those aspects of the business where the direct benefit can be measured in terms of cash. While this can be a good thing at the beginning of an EMS, when early savings from diagnosing inefficiencies in energy and waste disposal are picked up, as the system matures it will become increasingly difficult to secure management support for those projects where, although there will be a financial break-even point, the main reason for action is to reduce impacts on the environment. In phased implementation schemes, limited commitment on behalf of management may make the later stages more difficult to complete than the early ones as interest wanes and benefits may seem less tangible.

We want to improve/maintain our image. Lastly, those who are involved for the sake of the image or potential marketability of good environmental management are in for some surprises. As with most things related to the image of an organization, improved environmental performance is fast becoming something that the marketplace expects, rather than is impressed by. It may contribute something to an overall impression of an organization in the marketplace, but it will be largely unquantifiable. Thus if managers are expecting bulging order books, full to the brim of new work simply because they are managing their environmental impacts, they may well be in for a disappointment. Again, this may not show in the early support that an EMS is lent, but as the system becomes part of everyday management, it is likely that managers will begin to question the point of having one at all.

Obviously, some individuals can be motivated by more than one reason, but each one is subtly different from the next. In any individual, one reason will predominate and will in turn affect the way in which the commitment to an EMS, or even progress towards one, is expressed and carried out. It is highly likely that in your own organization the same prevailing reasons will be quoted. More importantly, if the implications of any one of them are not taken into account prior to installing an EMS, then the reader of this book may end up presiding over an incomplete management system that adds no real value to the way the organization carries out its function.

To avoid that situation, commitment must come from the very top of the organization. If the answer from the board to the question 'Do we want to do this?' is not a collective, considered and whole-hearted 'Yes', then be aware of the problems this will pose to successful implementation. After all it is only the most senior management who can and will:

- resolve cross-functional disputes;
- provide adequate resources;
- incorporate environmental concerns at a strategic level; and
- ensure a consistent approach to environmental management at all levels.

Knowing the nature of the commitment in an organization can also contribute to the 'maintenance of objective', a fundamental principle of complex operations and projects. Understanding the ultimate goals of a project is the only way a manager can remain flexible to the changing circumstances during the implementation process. Such clarity will also allow decisions on the system's future development to be focused on the central requirements and real issues. It is the best way to avoid the spread of needless paperwork and the creation of roles that simply preserve the system's original structure, ultimately to the detriment of effective management. This is especially true over the extended time-scales that can potentially be presented by phased implementation. In such schemes, every stage completed means that top management may wish to revisit the level of their commitment before embarking on the next one.

However often one has to do it, securing commitment is not as difficult as it may appear. Take a formal approach and ensure that decisions are recorded to avoid any later disputes. The basic advice is to remember that senior managers need to be convinced of three things in terms of implementing an EMS either in whole or part:

1 There is a problem or set of issues that can be solved by environmental management.

2 It is right for the company.

3 The benefits outweigh any downside factors.

Like most people, they will need to be convinced on a continuing basis. Under the remaining sub-headings, this section discusses some of the strategic issues that will help win and sustain that top management support. The following chapter will then help you in gathering your information so that an effective report on your current position can be compiled. Such a report may be the most persuasive tool available to you in gaining top management commitment.

Do you appreciate the role and influence of stakeholders?

Corporate governance (in short, whoever has control or can strongly influence organizations) is at the heart of the strategic issues surrounding environmental management. There is a huge number of theories concerning how to make a company more responsive to the needs of the many 'stakeholders' that it serves. A list of such stakeholders is provided in Table 2.1. Before you read on, examine the list and try to characterize more precisely the groups mentioned and what their expectations of your company might be. Also make notes on what you and your top management think of them. Even writing 'don't know' will show where you lack data.

For many organizations and their management, the expectations of parties other than investors and shareholders have often been regarded as being of secondary importance.

Table 2.1 Stakeholder review form

Stakeholder type	Applicable?	Expectations of organization	Evidence
Regulators			
Employees			
Unions			
Shareholders			
Bank			
Insurance company			
Neighbours/Local community			
Pressure groups			
Local authority			
Customers			
Suppliers/Contractors			
Other?			

However, current theories of management practice and broader interpretations of company regulation now recognize the shortcomings of being entirely profit oriented. Profit as a motivating force in business obviously has specific advantages – money can be shorthand for anything in a material society – but when the material society itself is changing, too tight a focus on ultimate profitability can blind an organization to other threats and opportunities. Current markets respond not simply to profitability alone (and it would be unrealistic to think that money does not still matter to most people) but to a complex interaction that takes into account how the profit is made.

Later in this section (see 'Do you understand how to define the issues?' below), we will talk about strategic level decision making, and the perceptual necessities and short-comings involved. In terms of the market served by your organization, it is important to realize that the stakeholders that make up your market may also be giving mixed signals about their attitude towards developing issues. One manager said that trying to get a coherent message on some issues was like listening to a sports commentary given by some-one suffering from extreme short-sightedness: the responses were long on atmosphere, but short on detail.

Such 'market astigmatism' is a key point to remember. Up until the early 1990s, business organizations have been able to exploit freely areas that have hitherto been invisible or at least barely discernible. The environment has been out of the sight and mind of both organizations and the market they serve. Much work is being done in bringing environmental costs within the bounds of the financial systems that already exist, thus making the impacts quantifiable and 'visible'. Though the relative success of this tactic has varied, the continuing attempts are at least recognition that something vital has been missing from strategic business calculations and decisions in the past. No company or organization can exist in a vacuum, free from the influence of the stakeholders that it seeks to serve. Understanding the range of pressures that apply to your own organization in terms of the environmental impacts that your organization generates is key to designing an effective EMS and will guide your organization through the strategic level decisions it will have to make.

Do you understand how to define the issues?

At the strategic level, management systems both formal and informal are simple things. They are mechanisms for moving information around inside an organization in order to facilitate a better quality of decision making. Interestingly, they also have a secondary, but nonetheless important, function; namely, acting as a divining rod for the type of decisions that the organization will have to face in the near future. For this reason, to reduce an EMS to an 'operations only' approach (as often happens to health and safety issues) opens top management to the possibility of being 'blind-sided' by important strategic issues.

Take some time out from reading this book to consider what you think the major environmental management issues are for your company. As an employee and a member of society, you are currently in as good a position as anyone to hazard a guess as to what the major issues might be. To help you structure your thoughts, simply write down your organization's activities, services and products and try to think of what environmental aspects there may be associated with those elements. It doesn't have to be an exhaustive list, or even a technically informed one. Your first impressions are, however, worthwhile recording for comparative purposes after you have carried out the exercises in Chapter 3. Table 2.2 provides a sample issue list and a worked example to help you. You may even find it useful at some stage to ask others in your company or organization to fill out a form like this, just to see how perceptions differ according to their function and perspective. Keep your own original list on one side, and use it as an aide-mémoire until you've read Chapter 3.

Table 2.2 Sample issue list

Activity service or product	Impact on air	Impact on water	Impact on land	Energy issues	Waste issues	Noise, vibration and odour issues	Other? (eg biodiversity, social, etc)
Transport of products and raw materials	Traffic emissions	Potential diesel spillage into sewers on site	Land use for vehicle park required	Use of fossil fuels	Waste oil, end of life tyres, etc?	Increased noise in nearby residential areas	Safety issues for local community

At this stage, you may not have enough knowledge to be able to define accurately what the environmental issues are that your organization has to deal with. Many phased implementation schemes begin with a group exercise that is little more than an organized version of the same 'brainstorming' technique. However, what you will experience by doing exercises like this as well as the ones in the rest of this book is a gradual expansion of your knowledge; it is worth remembering that many others in your organization will have to go through the same process. By using these exercises to blaze a trail, you'll be able to look back, see how you got where you are, and ease the path for others when necessary.

Many environmental issues are far from clear cut. In some areas, the perceptual debate may never cease; for instance, humankind's contribution to the effect of global warming. Climate change is officially recognized, but opinions differ as to the extent that industrial impacts play or whether it is part of a larger cycle of climate fluctuation, a bigger pattern that we can sense but cannot yet see in full. The political and liability issues that recognition of human intervention in climate change implies are also holding back further action and research. When issues hover on the edge of knowledge in this fashion, half in and half out of the light, they have the same disturbing tendency on business decision making. Prior to making the journey that crosses the dividing line between known and unknown, any issue will cause a wavefront of perturbation, not just in an organization but in the market and the wider business ecology to which it is linked. The earlier those disturbances are felt, the earlier they are taken into account during decision making exercises and, in turn, the earlier better quality decisions are made.

Repeated perturbation around an issue eventually leads to its transformation. Imagine a tide coming in up the beach; every wave comes in a little bit further, and each time it recedes, it doesn't go out quite so far. Each wave is like a ripple of perturbation that will eventually lead to the transformation of the beach into a different state, where it is covered with water. All of us recognize the final transformation easily enough and talk of an idea having reached 'critical mass', or of a concept whose 'time has come', or of something that is now 'a fact of business life'. We are distinctly less comfortable with continued perturbation.

Managers in particular are loath merely to watch the tide come in and many attempt to leap-frog the phase altogether. For them, it is infinitely preferable to adopt a constantly revolutionary approach to new strategic issues, rather than an evolutionary one, preferring to go straight for transformation. However, quantum leaps in organizational knowledge caused by such forced non-spontaneous transformations are rare, and when they do occur, frequently render everything that has gone before useless. Anyone who has had experience of what happens after take-overs, mergers and management buy-outs can bear witness to this.

A good management system will be mapping any strategically disturbing wavefronts. The mapping will allow a less painful easing into change, rather than the step change model of 'making forward progress, hitting a brick wall, climbing up the wall, making forward progress, hitting a brick wall . . .' in an endless cycle of acceleration, painful impact and upward struggle. A good EMS will open the door to increasingly flexible management, based on far-sighted and balanced appraisal of future developments. And if it can happen in one area of management such as environmental aspects of a business, it can happen elsewhere.

How does the supply chain affect your EMS?

There are lots of similarities between the supply chain and an ecological food chain; namely, your relative health and your mode of survival depend on where you are in the chain. As with most things in life, there is a degree of simplicity about this situation if you are at either end of any one chain. The difficult part for most of us is that we are often somewhere in the middle.

Interestingly, in environmental terms, it is those closest to the customer who have noticed the requirement to address environmental issues properly and in a systematic manner. Retailers (in the food and timber areas especially) are the last link in a chain before the handover to the customer, and it is they that bear the brunt of market indecision, fears and anger when things go wrong. It is hardly surprising then that they should turn to those suppliers further back in the chain and express a sincere and heartfelt desire to manage such an area of consumer concern as the environment.

Elsewhere, banks and insurance companies have an equally close relationship with their customers and their suppliers simultaneously. The financial services sector use their acumen to 'produce' more money from investment and, at the same time, sell their services directly to customers. As a service industry, at times it is hard to discern anything like a supply chain at all, and the convenient dividing line between customer and supplier is not so easily drawn. As a result, they are in a position where they must develop robust policies to protect themselves from the worst of the potential financial penalties that may arise from environmental pollution (and thus threaten the continued long-term performance of an investment) and at the same time, encourage their customers to take environmental management seriously as a hedge against future liabilities.

It is not only the service sector that has unique challenges lying ahead in terms of environmental management. No two organizations, even if they are in the same industrial or commercial sector, have the same relationships with their suppliers and customers. Each situation is unique. As a result, each solution to a specific environmental impact may be different, linked as it is to those further up and down the chain. This may lead to unusual cooperative ventures, more akin to the symbiotic arrangements found in ecological systems than in business development. Such relationships are also being fundamentally changed for the better due to the introduction of EMS phased implementation schemes. Some schemes are flexible enough to be supported by larger corporates, who commit resources to make training and consultancy available through a supply chain initiative. These can be underpinned by either second or third party audits of the EMS elements as they are put in place. There is even the possibility that more customers (possibly seeking continual improvement under their own ISO 14001 or EMAS registration) will undertake a risk assessment of their supply chains and use the outputs in conjunction with a phased implementation scheme to tailor EMS element requirements for individual suppliers. (For more information on this see Appendix I.)

As an integral part of the establishment of a formal EMS, an organization must get a sense of where it is in its own unique supply/food chain, and understand the relationships between the business environment and its own impact on the global environment. This will bring a new level of understanding and potentiality to the organization's management, and may well lead to greater use of mutual benefits within the chain to identify areas of commercial and environmental improvement. Perhaps more to the point, failure to understand these relationships could seriously hamper future business development compared to competitors who are already seeking them.

Is external recognition appropriate for your EMS?
The international standard ISO 14001, the European Union's EMAS regulation and some national phased implementation concepts like the UK's Acorn/BS 8555 scheme have been

designed in such a way that anyone meeting the requirements of the schemes can seek external recognition through certification and registration of their management systems by a third party. These third parties are so called because they are independent of both the organization running the EMS and their customers. They usually take the form of accredited certification bodies, in the case of ISO 14001, independent verifiers in the case of EMAS or accredited inspection bodies in the case of Acorn. For brevity, we shall refer to them all under the term 'certification'. There is a degree of overlap between all the schemes and they are certainly compatible. Still, care should be taken if an organization is planning to utilize all the schemes over time (see Appendix I – Understanding the Requirements of EMS Standards).

Organizations need to be clear on the value of independent certification to the schemes, in particular, relating the worth of attaining registration to their specific circumstances. Obviously, there are a range of circumstances where the choice may be limited:

- where certification is part of a larger corporate policy;
- where certification is demanded directly by clients; or
- where certification is demanded directly by regulators.

If none of the above circumstances apply, organizations should take into account that many certification bodies operate in an open market, selling their services in competition with one another. As such they are seeking to maximize their own financial performance and it is in their interest to sell the benefits of a certified EMS. On the other hand, it is down to the individual organizations themselves to assess what benefits accrue from having their EMS certified over and above simply operating an EMS.

What external recognition can do is demonstrate to a wide variety of stakeholders that the organization is committed to managing their environmental impacts. This may be useful if operating at international or European level, if your organization is responding to enquiries from potential clients, regulators or insurance companies or if you want to provide a valuable incentive to your own operational staff. The key to assessing the value of certification to your organization is to carry out your own survey of the potential benefits. Below are listed five of the main considerations; read these in conjunction with Appendix II – Third Party Auditing, where the certification process is discussed in more depth.

1   The market
    (a)   Are any of your customers or other stakeholders demanding certification of you?
    (b)   Are they influential?
    (c)   Will any of them make such demands in the future?
2   The costs
    (a)   Is there local/regional/national funding available?
    (b)   Is there a phased implementation scheme available? Does it have local or customer support available?
    (c)   Are you already registered to ISO 9000?

(d) Have you got quotes from a variety of certification/verification bodies?

(e) Have you investigated how any existing or future certificates could be structured?

(f) Are you sure you know what the maintenance/surveillance costs will be?

3 The benefits

(a) Have you quantified the benefits of certification/verification, over simple EMS operation?

(b) Would gradual or partial progress through a phased implementation scheme suit you better?

(c) Will external recognition add more impetus to the implementation?

4 The alternatives (ISO 14001 focus)

(a) Have you discussed alternatives to certification with your stakeholders?

(b) Would phased implementation and a simplified EMS suffice?

(c) Are second party (customer/supplier) audits possible/desirable?

(d) Is self-declaration against ISO 14001 appropriate?

5 The choice (EMAS focus)

(a) Is there a corporate preference for EMAS?

(b) Do you serve a mainly European market?

(c) Is making a public statement of your environmental performance appropriate? (EMAS requirement)

(d) Are you a local authority? (EMAS is better structured for local government.)

Narratives

Smallco

The quality manager's report went well, especially the references to cost savings which might be achieved. The offer of help from one of the suppliers and the benefits of a gradual, phased implementation of an EMS make good business sense. The MD has now designated Smallco's Operations Manager as the environmental management representative as he has previous experience of EMS. In practice the hard-pressed Ops Manager feels he has drawn the short straw. After only five months in the job, there is still much of his original brief to complete, including major changes to the production schedules and the processes themselves.

The Ops Manager feels unable to say no. However, there are many ways to say yes, and his experience as a environment team member in his last job indicates teamwork coupled with senior management commitment is the only effective way to achieve success. Anyhow, much of the work to re-engineer production schedules and processes is already revealing opportunities for improvement. The Ops Manager sets himself five key tasks:

1 To further investigate phased implementation of an EMS based around Acorn/BS 8555.

2 To map out a draft implementation plan (cunningly derived from the one he has from his last job, adjusted for Smallco's situation and the well received idea of a phased implementation).

3 To begin discussions with management colleagues and the supplier who offered help to sound them out for practical support if it comes to an EMS implementation programme and to find out what will make sense for them.

4 To gather some ideas for cost savings and guesstimate returns and present these at the first strategic steering group meeting. The quality manager tabled 20 ideas at the earlier meeting.

5 To present a skeletal plan to the MD which also involves him chairing a 'strategic steering group' – vital so the Ops Manager can remain neutral. This group would only meet quarterly and as part of regular meetings already taking place. Note the clever use of terminology. The actual work would be done by an action team in each functional area of Smallco, coordinated by the Ops Manager. In this way the senior management would be engaged and over time buy-in would be achieved while others do the legwork.

The Ops Manager has set himself the target of four months to complete these tasks. The only slightly hidden agenda is to ensure that costs and benefits are communicated to the MD. The next management meeting is crucial.

BIG Inc

At the first board meeting since the appointment of the project manager, the directors review a schedule of the certification project. It includes an allocation of some five months to allow for a detailed initial environmental review (IER). The directors are not convinced that the IER should take that long, as they think the company doesn't have many environmental impacts. They also think that exclusion of the European offices and home-based staff will reduce the need for such an extended review still further. During discussion, it also comes out that 50 per cent of the board are sceptical of the ultimate value of registration, having had a bad experience with the company's registration to ISO 9000 which required copious documentation. The same directors thought that ISO 14001 would simply require cloning the documentation.

The project manager, realizing that the board was not completely aware of the level of commitment required, outlines the differences between a QMS and an EMS and insists on her original estimate. She explains that their existing ISO 9000 certifier (the nominal choice for EMS registration), when asked about the scope of the system, told her the European offices and home-based staff would have to be included to ensure the whole organization was covered by the review. To avoid further commitment problems, the project manager secured an early designation of a board member as the environmental management representative

(EMR), agreement to the submitted schedule and further agreement to detailed ongoing reports by the new director/EMR on the progress of the project. Longer term, she plans to get across the strategic potential of the results of the IER to the board.

Trade secrets

- Start securing commitment from top management. Read the next chapter and prepare a report for them that covers:
  - all the methods that you currently use to address all of the strategic issues above;
  - the scope of the initial environmental review;
  - potential benefits of improved environmental performance;
  - the allocation of responsibility for environmental performance at board level; and
  - conclusions/suggested actions.
- Submit the report, make a presentation and aim to get commitment to carry out the initial environmental review only (decisions about EMS installation and any external recognition will only come after the review).
- Ensure that the potentially emotional nature of the issues involved in environmental management is not allowed to derail what is basically a business development.
- Staff have been known to suffer from 'initiative fatigue' if there are too many business projects going on at once. Try to get a prioritization of all management initiatives to avoid competition. Can one of them be postponed, or given a longer deadline?
- It is not worth considering 'cloning' a paper-heavy QMS in order to provide the basis of your new EMS. Emphasize any similarity with an established ISO 9000 system by all means, but only when it is in the organization's interest (ie to avoid duplication, etc)
- Carry out an exercise with the appropriate management team to define and agree the exact success criteria for the EMS implementation, including qualitative indicators of success, over and above time and resource constraints.
- Do not start the EMS implementation without a written list of the agreed success criteria, or it will be tantamount to setting sail without any charts.
- Competencies and skills may already exist within the workforce, but do not expect past experience of environmental issues to be an obvious part of training and competency records. Fine judgement is required to ascertain whether a personal interest is supported by any useful practical experience.
- 'Get the certificate' is too narrow a success criterion to bring lasting benefits of an EMS to the company.

? Things to think about

- The future of environmental management in an organization needs to be thought about from the very beginning. If an EMS is associated too closely with health and safety or quality issues in the minds of top management, it will be hard to get environmental issues considered as anything else other than in operational terms. All management systems share similar delivery mechanisms, but not strategic perspectives.

- The environment is aligned more closely with economic and social performance in terms of an organization's sustainability, so it is worth emphasizing why an environmental perspective has serious ongoing implications for the organization, no matter how small. Sustainability (and even Corporate Social Responsibility) is an agenda that is developing for everyone, not just large commercial businesses.

Box 6.1 Best Foot Forward environmental policy

Best Foot Forward is committed to reducing its impact on the environment. We take our responsibilities seriously and pursue a policy of environmental best practice with clear policy aims and objectives.

Policy aims

- To include environmental considerations in daily and project activities
- To conduct an ongoing company resource flow analysis and ecological footprint
- To implement an environmentally responsible purchasing policy
- To reduce, recycle and reuse waste on the premises and at BFF events
- To maximise energy efficiency and reduce greenhouse gas emissions
- To minimise and control the use of water on the premises

Objectives

Energy

BFF aims to maximise energy efficiency and reduce emissions by:

- minimising and monitoring its total energy consumption
- purchasing, where possible, renewable energy

3

The Initial Environmental Review: Where Are We Now?

An IER provides the foundation for your EMS. It highlights your environmental aspects, applicable legislation, what is already in place and any learning from previous incidents. The knowledge and understanding you will gain from carrying out a review will ensure that the other key elements of the EMS – policy, objectives and targets and the management programmes that bring the EMS to life – are based on a sound understanding of your situation and reflect this reality.

✓ **ISO EMAS quick check**

Area of EMS	ISO 14001	EMAS
Initial environmental review	A.1	Annex VII & Article 3(d)

📖 **Chapter executive summary**

Move to the next chapter when you can answer all of the following questions in ways which make sense to you and your organization:

What is involved in scoping and preparation?

What are the specific IER communication issues?

What information needs to be brought together?

What level of analysis is required?

SETTING UP

🔖 **You are here**

... through techniques and tools for doing so are rapidly developing. The strategy which is embodied in the environmental ... needs to reflect these changing factors in the business environment and be flexible ... to respond positively ... as ... business benefits it is well worth getting hold of a wide variety of policies from other companies – not just those from your own sector – and see if there is anything you like about them and if there are any elements which might help in your own policy drafting work.

ACTING
14 Management Review

SETTING UP
1 Introduction
2 Commitment
3 Initial Review

PLANNING
4 Aspects and Impacts
5 Legal and Other
6 Policy
7 Objectives and Targets
8 EMP

CHECKING
13 EMS Auditing

DOING
9 Communication
10 Competence, Training and Awareness
11 Documents and Control
12 Operational Controls

14 Management Review
ACTING

Drafting the policy

The environmental policy is a cornerstone of the EMS. At this stage you should be ready to begin to assemble a first draft of the policy. It is well worth preparing a first draft for internal consumption only. Make sure you mark all copies up as 'Draft only' and 'Confidential' – if you don't they have a habit of leaking outside before you are ready. You will have assembled a team of trusted colleagues helping you to develop the EMS. They will have been closely involved in the IER and setting of objectives and targets which underpins the policy commitments, so make use of their expertise for critical comments on this first draft. It is worthwhile adding a couple of neutral readers to the list of critics, for example, a trusted colleague from another organization who will be able to take a disinterested, external view of the draft. Their observations can be of great assistance in developing a policy which will make sense to a wider audience.

With this expert input, redraft the policy to produce a presentation draft for discussion with senior management. It may help to discuss this draft policy alongside your emerging objectives and targets for the EMS. Remember, the policy provides the framework for setting objectives and targets so the final sign off will come when all are satisfied that the policy and the objectives and targets are working in tandem. Finally, produce the first version of the organization's environmental policy and make sure it is a controlled document and signed off by senior management.

What needs to be done with it in the context of the EMS?

There are several things to do with the new environmental policy. Some, such as making it available to the public, are requirements of the standards. Other actions are fundamental to making the policy an instrument of the EMS. Use the checklist below and consider each point in turn.

• Make sure you send a copy to everyone who helped in the consultation process leading to the drafting of the environmental policy.
• Distribute a copy of the policy statement with a suitable endorsement from senior management to all employees. This could be by paper or electronic means.
• Attach copies of the policy to workplace notice-boards and in staff handbooks.
• Publish the policy in company reports and other relevant communications.

📋 **Toolkit requirements**

To work through this chapter you will need the following knowledge and skills:

Knowledge

- basic environmental business issues for business;
- your organization and what it does;
- any business or other plans and policies;
- key drivers for environmental management;
- top management commitment(s) including available budget(s);
- roles and responsibilities in the organization;
- stakeholders – internal and external.

Skills

- information collection, handling and interpretation;
- communication skills – oral and written;
- analytical skills;
- information gathering and handling;
- interpersonal and observational skills, especially interview and listening;
- note taking;
- managing people.

Context

As the critical first step on the road to effective environmental management, the IER is a systematic examination of past, present and projected environmental position and performance. Put simply, it is a stock take which will signpost the way ahead for the EMS. It will need to cover all aspects of the organization, from systems to staff. The information revealed will become the baseline from which to measure future progress and to build the EMS itself. It does this by providing a thorough insight of the business position, including any exposure and liability issues, against which meaningful objectives and targets can be set. The review will identify organizational strengths and weaknesses as well as any risks or threats. Most importantly taking stock will identify the opportunities to integrate proposed environmental actions with best business practice.

It is a truism in management that if you can't measure it then you can't manage it. Carrying out an IER will enable a plan of action to be set out with knowledge and understanding of your organizational situation and, as a result, confidence that what you are doing will make sense for the business and the environment. Your objectives and targets and environmental management programmes (EMPs) will germinate from the intelligence developed by the review process and ensure business-based priorities and targets are clearly identified and realized. An IER will ask:

- What are the key issues?
- What are the opportunities in terms of standards and benefits that could be achieved?
- What is required to ensure 'legal compliance plus'?
- What are others doing and how can we learn from this good practice?

Your paramount objective is to carry out a thorough IER of your business. The focus associated with each part of the review will vary depending on organizational needs; the issues remain the same. The outcome of the IER work will enable you to answer with confidence any questions relating to your management of environmental issues in the organization and to have developed the foundation for an effective EMS.

We have found that implementers spend at least 40 per cent of their total EMS project time on the IER alone, and the more detailed the work done, the less time is wasted at a later date. Skimping on the IER is false economy in terms of time. The review itself can be separated into three distinct steps: (i) scoping and preparation; (ii) collecting the data; and (iii) making sense of it.

Tasks

What is involved in scoping and preparation?

The earlier chapters provide persuasive arguments and a rationale for taking environmental management seriously. We could argue that the environmental debate is over and that environmental management or 'being green' is no longer fashionable or glamorous, it is simply a fact of business life. Speculative questions like 'Will the environment affect the business?' are missing the point. It is not a question of will it, but how much and how soon? Conducting a review will provide a planned management response to these arguments.

For those looking for external certification, there is of course the other substantive argument. In one way or another, both of the management standards – ISO 14001 and EMAS – and even phased EMS implementation schemes require an IER. EMAS, the Eco Management and Audit Scheme, clearly states that an environmental review will take place and that it 'shall mean an initial comprehensive analysis of the environmental issues, impact and performance related to the activities at a site'. It is also a requirement of EMAS (EMAS Regulation, Article 3(d)) that the environmental review is externally verified and deemed to be technically satisfactory. In the same way, phased implementation schemes, like those built around BS 8555, have a baseline review that is a recognized and externally inspected part of the very first phase of work. Although not strictly a requirement of ISO 14001, Annex A.1 advises that an IER should take place in an organization with no existing EMS. In our experience it is advice that is hard to ignore. If you haven't identified where you are starting from you can hardly approach your EMS in a systematic manner.

Given that the key purpose of an EMS is to reduce environmental impacts of the organization (in ways which make business sense), then developing a thorough understanding of environmental aspects and impacts of the organization through a review process is paramount to ensure that the EMS will function effectively. Setting the scope of the review will serve to identify current practices and any learning from past experiences. The direct results will include:

- *Compliance plus:* the confidence that your business is complying with existing regulations relating to your environmental aspects and any other requirements **and** will be able to embrace future legislation with minimum disruption to your business.
- *Cost savings:* the review will reveal areas of the business where you can save money and so sustain a competitive edge. Early wins also lubricate implementation.
- *Aspects and impacts:* developing an understanding of your greatest environmental risks and reduce any associated threats. This understanding is fundamental to establish meaningful objectives and targets for the business which will then be implemented through the management programmes for the EMS.
- *A realistic environmental policy:* the environmental review will provide the information needed to draft a meaningful organizational policy.
- *Implementation:* the information to establish a business-focused set of actions to move from the review towards a business-focused EMS.

Jot down half a dozen key benefits that you need from your IER. Keep this list and see if your priorities are reflected in your final objectives and targets.

The IER will allow you to focus on those areas of the business that may need consideration and those areas which represent the best opportunities. The checklists provided later in this chapter provide one mechanism to guide the review process and can be used as a starting point for your review. Carrying out the review on your own will not deliver the results you need. Without ownership and involvement from colleagues, a cloistered approach to the IER is also a certain way to make implementation of its findings hard work. Teamwork is the key to success. The team you put together can be drawn from inside and outside the business and individuals can be involved in the review process as much as required. In IER terms there are some golden rules of engagement, so make sure:

Senior management are involved and committed at all stages in the review process. It is, after all, their time and money you will be using and their support that will be needed if an effective EMS is to be developed and sustained. Remember that at this stage their commitment is to a full IER. You are not asking them to commit to a course of action beyond an examination of the findings at this stage.

You have access to any specialist skills you are likely to need. This is an important part of scoping the review and determining not just what you will do but how you will do it. These skills may be from within the organization and may include specialist or technical staff or from outside the organization and based on developing or established working relationships with the external environmental professionals. The regulators themselves may fall into this specialist category.

Staff know what you are doing and why you are doing it. As we will discuss in Chapter 9, staff commitment or the lack of it is the greatest opportunity and threat to the viability of any project. Building an EMS is no different and may well raise new and diverse challenges to those whose involvement is essential for its implementation.

Those involved understand what is involved in the work and can do what you want them to do. This might sound like a truism, but if people are not sure what is expected of them then they will do what they think makes best sense. This may not be the sense you actually had in mind!

You need to decide the depth and breadth of your IER. Remember, ISO 14001: 2004 requires the organization to define and document the scope of its EMS. Is it to cover all organizational functions in every last detail? Will the first step in the review process be to take an overview of the whole business and then focus on identified priority areas? Will you home in on one site of a multi-site business and use this learning to apply to other similar sites? Will you carry out the review in an incremental way by focusing on one or two areas first then building the complete picture step by step and issue by issue over a pre-determined time period?

Assuming that achieving one or both of the formal standards or a related scheme is an ultimate goal, you will need to satisfy their key criteria. It may be that achieving conformity with the standards is a longer term goal, in which case the review activity may focus on more immediate business needs than standards requirements. Whatever you do there are several basic considerations and associated questions to ask yourself:

1   *Areas or activities – what's the scope of the EMS?* What areas or activities am I going to assess? Areas might be functional areas of the business, for example, a workshop or office suite. Activities might be transportation or raw materials sourcing and purchase.

2   *The information.* What information is needed, what information already exists, where is it, who keeps it and how best can it be assembled in the context of the environmental review?

3   *Timetables.* Given the work that will need to be done and the people who will need to be involved, what is a realistic time-scale to carry out the review?

4   *Resources.* What resources are available? As well as the more obvious considerations of time and money, resources might include internal expertise and records and relationships with external sources of guidance.

5   *Communication.* How will effective communication be realized? This will include the review team, key staff with knowledge and information that needs to be considered and staff in general.

Table 3.1 provides some guidance on building a project outline for your review and prioritizing issues which you feel need attention in your organization.

What are the specific IER communication issues?

In terms of the IER, you are now in a position to move forward and begin to collect the hard data for your organization. Before continuing, now is the time to hold a briefing meeting with key internal staff and formalize the communication process with staff in general. Even in the smallest organizations staff commitment and involvement will help ensure success. Throughout the development of an EMS, it is of paramount importance to communicate with your colleagues. After all, it is they who will make or break the system. People will only give their best if they fully understand the decisions that affect them, how and why these decisions arose in the first place and how their contribution can and will make a difference. Developing a sound understanding of staff role is a key component of effective EMS implementation and management. At this stage in the IER preparation, it makes good sense to engage colleagues and let them know what is going on.

A first briefing session might well be part of the agenda of a general management meeting. It is worthwhile providing a focus for your slot to set the scene, report on progress so far and indicate the next steps in the process. A short briefing paper, perhaps only a side or two of paper, could cover the following:

* gaining support/obtaining buy-in from others.

Table 3.1 Project outline for initial environmental review

Area of business	Score (*)	What aspects?	When?	Who do I need to involve? Internal	External
Management issues and awareness					
Legislation					
Marketplace					
Distribution and transport					
Waste and discharges					
Paper and packaging					
Site management/Good housekeeping					
Planning/Development/Land issues					
Product					
Process					
Raw materials					
Hazardous materials					
Water					
Energy and fuels					
Stakeholders					

(*) Score the environmental issues associated with each 'area of business'. For each use a scoring system to derive an overall score for each area. Think about three assessment criteria and give each a score of 3 for a high rating or 1 for a low rating. Once you have given a rating, multiply each rated criterion together to give an overall score and enter this in the chart above.

An example is given below for distribution transport:

Environmental hazard (2); Likelihood of the problem arising (3); Size of the problem (2).

Overall score = $2 \times 3 \times 2 = 12$

You may, of course need to sub-divide each area of business into a number of component parts and prioritize each component. This is dealt with in the sections covering issues and information below.

SETTING UP

- Why are we carrying out an IER?
- What is planned – perhaps include your extended version of Table 3.1.
- What benefits will be gained from the work? Your interpretation and application of the first two chapters of this book will provide input here.
- The practicalities of the review in terms of who does what and what will be expected of employees at each stage in the process.

Your report is also a good opportunity to take feedback and allay any concerns colleagues might have about the process. Often some constructive ideas will spring from the discussions which will help fine tune the process and show that you are prepared to listen and take comments on board.

What information needs to be brought together?

Having set your strategy it is now time to do the work. Think about your plan for the IER and the information gathering process. The summary chart (Table 3.1) you used in the scoping and preparation will help by providing the introduction for the review process. These 15 issue areas can now be expanded to provide a more detailed analysis of the environmental position of the business.

For each issue area, a series of starter questions are included below to begin the process of collecting information and hard data. The questions raised are examples only and you will certainly need to add further questions based on your own situation and the priorities for your organization. In reality each question will raise many supplementary questions which together will provide a comprehensive picture of your environmental situation. Some areas will require extending your work, perhaps by seeking guidance and advice from external sources such as the regulators. As you begin the data collection and analysis of the issues in your IER you will no doubt extend and focus the scope of your own work. This dynamic fine tuning is typical of most projects. Whatever you do in terms of detail, keep the following primary thoughts in mind:

- What are the environmental aspects for each business function?
- Who actually has the responsibility for the areas we are looking at?
- What is the scale of any risk identified? (Refer to your scoring system in Table 3.1.)
- Is there a potential 'cost' to the business?
- What is the cost of doing nothing?
- Talk to people who are actually involved – what do they think and what advice can they provide?
- What procedures are actually in place, could be in place and should be in place?
- Depending on your relationship with your regulators, ask for their help and advice – many do want to encourage and help businesses to achieve best practice. Bear in mind, however, that national and regional differences in approach will exist among regulators, so judgement is called for.
- Try and find out what others are doing, learn from their experiences and identify best practice for your sector.

Environmental performance evaluation

As you embark on your IER it is also well worth considering environmental performance evaluation or EPE for short. If you have opted for a phased implementation scheme like BS 8555 then this begins with establishing and developing environmental indicators for your organization.

EPE is quite simply an internal management process that provides information to enable management decisions relating to the organization's environmental performance. EPE was one of the original five main areas for an ISO 14000 series which had been set up by ISO Geneva in response to the Rio Earth Summit in 1992. The detailed methodology for EPE is set out in *ISO 14031: Environmental Performance Evaluation.*

EPE involves collecting information to measure how effectively you are managing your environmental aspects on an ongoing basis and can provide assurance in several areas including:

- A more detailed understanding of your actual impacts on the environment.
- Data to benchmark environmental performance.
- Highlight areas for improvement.
- Understanding whether objectives and targets are being met.
- That legislation relating to significant aspects is being addressed.
- Adequate EMS resource allocation.
- Data to allow reporting with confidence and internal and external communications.

Like an EMS itself, EPE follows the Plan-Do-Check-Act (PDCA) model:

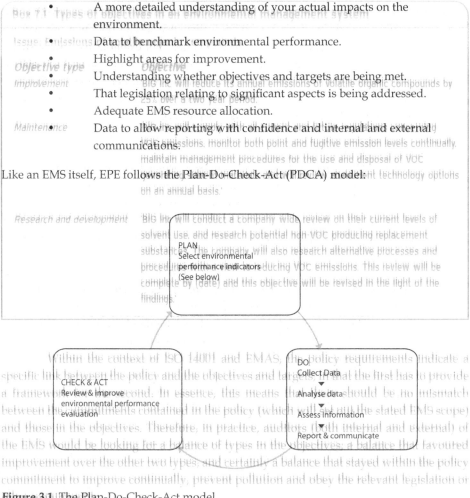

Figure 3.1 The Plan-Do-Check-Act model

Environmental performance indicators (EPIs) come in three varieties:

1 management performance indicators (MPIs);
2 operational performance indicators (OPIs);
3 environmental condition indicators (ECIs).

By way of example, consider how Shelox Brewery has used EPIs:

Table 3.2 Use of EPIs by Shelox Brewery

Objective	Programme	Performance indicators	Type
Water conservation	'H$_2$Zero'	Water use per unit of production	OPI
Waste water treatment	On site treatment plant	Volumes of effluent processed	OPI
		Energy used by plant	OPI
		Quality of discharge water to river	ECI
Employee training	Operator and awareness training	Employees trained and competent	MPI
Community relations	External relations	Newsletter	MPI
		Positive press coverage	
Minimize resource use – timber pallets	Resources and waste	Pallets scrapped	OPI
		Reduced timber use	ECI

At this stage, it is a good idea to think back to any earlier review team or other staff briefing meetings. You will have come up with a preliminary ranking of the 15 environmental issue areas for your organization. It is worth reviewing how each of these issues relates to your business activities and how they are related to each other. For example, the way you package finished goods may well determine how they are stored and transported to the customer. This in turn will determine transport infrastructure and costs and reveal environmental components of each activity.

Below, for each of the 15 issues, you'll find a quick résumé and some key questions to start you on your information gathering which will provide the basis for the IER. The 'Possible actions' boxes will need some extra paper and are for you to outline your early thoughts on what you might do to manage these issues.

Issue 1 Management issues and awareness

By developing a comprehensive understanding of what and who make the business tick, this 'issue area' provides the big picture of the organization and the context for the EMS. Some key questions:

1.1 Are any management procedures/systems in place with respect to environmental management?

1.2 Does the business have an environmental policy or are there any environmental components in other policies such as health and safety or quality?

1.3 What training and staff involvement already takes place? What opportunities exist to weave an environmental strand into this?

1.4 What is the senior management commitment to environmental management? (See Chapter 2.)

1.5 What is the culture of management and how would environmental management fit with this culture? (See the Introduction.)

Possible actions:

_____

_____

Issue 2 Legislation and other requirements

Complying with the law is a major obligation for any organization. Legislation is continuously changing and the key to success for any organization is to anticipate change and strive to modify operational practices to stay ahead of any changes to legal or other requirements.

2.1 What legislation applies to the business and how is it implemented?

2.2 What is the business response to meeting environmental legislation and other legislation with an environmental component?

2.3 How is legislation monitored?

2.4 What relationships exist with regulators?

2.5 How are these relationships managed?

2.6 How is legal compliance evaluated?

2.7 What 'other requirements' apply, for example, industry codes of practice or local requirements?

Possible actions:

_____

_____

Issue 3 Marketplace

3.1 What customer policies exist? How are these responded to?

3.2 Do suppliers and subcontractors know the business's supply standards/policy?

3.3   How do they respond? How are these relationships managed?

3.4   Is awareness raising/training part of the supplier/subcontractor relationship(s)?

Possible actions:

_____

_____

Issue 4 Distribution and transport

A necessity of business life and a significant impact on the environment. Your impacts might arise through your company car fleet, distribution vehicles or use of contract delivery services. Remember too, how you source your raw materials will have transport implications:

4.1   What is the business mix of distribution and transport?

4.2   How much is spent on distribution and transport by:

(a)   cars?

(b)   lorries?

(c)   contractors?

4.3   What management control measures are used?

4.4   How could distribution/transport costs be reduced?

4.5   Are vehicles regularly maintained and legal?

4.6   What driver training is provided?

4.7   How is the site managed (eg fuel storage, vehicle loading and washing, etc)?

Possible actions:

_____

Issue 5 Waste and discharges

Legislation increasingly requires that waste must be managed and you may have a direct responsibility for it. Remember, waste = cost; at least once to buy it in as raw materials and once to dispose of it. Similar legislative requirements will apply to the discharges produced as a result of your operations and health and safety aspects of waste handling by operatives. These too must be managed properly:

5.1   Where and what are the wastes/discharges for each business activity?

(a)   to air?

(b)   to water?

(c)   to land?

5.2 Who is responsible for them?

5.3 What is the cost to the business?

5.4 In terms of procedures, what:

    (a)   is in place?

    (b)   should be in place?

    (c)   could be in place?

5.5 What are the relationships with regulators?

5.6 How could wastes / discharges be reduced at source?

5.7 Are there any community issues with respect to:

    (a)   noise?

    (b)   nuisance?

    (c)   vibration?

    (d)   odour?

Possible actions:

Issue 6 Paper and packaging

This is an important issue. For example, the European Union has established a packaging directive with specific targets for recovery and recycling of packaging. Member States may have their own targets, but it doesn't stop there – globalization is making packaging a worldwide issue. Paper use is a fact of business life. Even the so called paperless office, promised just a few years ago, has failed to become a reality:

6.1 Where is most paper/packaging used?

6.2 What opportunities exist to reduce materials used?

6.3 What relationships with customers exist with respect to packaging?

6.4 What recovery liabilities exist?

6.5 Is packaging fit for reuse?

6.6 Are there plans for packaging minimization?

6.7 Are there markets for 'waste' packaging materials?

Possible actions:

Issue 7 Site management/Good housekeeping
An understanding of what activities take place on the site and how the site operates are important considerations. Site management offers many opportunities to minimize exposure to regulation and reduce environmental impacts. Just supposing a local environmental health official turned up tomorrow; would you feel confident about giving them a tour around your site?

7.1 How is the site used and what is kept on it?
7.2 Are storage methods and procedures appropriate for the materials being used?
7.3 How well is the site managed?
7.4 What records are kept and how are these used?
7.5 Are staff trained in procedures and handling of materials?
7.6 Is the site/premises tidy and well maintained?
7.7 What relationships exist with the regulator with respect to materials stored and handled?
7.8 Are there any geographical/physical issues or relevance?

Possible actions:
_____
_____

Policy
Issue 8 Planning/Zoning, development and land issues
Knowing what regulations apply and having a good knowledge of land use issues is important information. You may have inherited a site from another business or be planning to move to new premises:

8.1 What do you know about your site history? What has gone on before?
8.2 What planning/zoning or land use regulations apply? Local, regional and national?
8.3 Do you plan to develop your site? What will this mean in terms of land use and what are you required by law to tell the planning/zoning authorities?
8.4 Do you plan to move site? What will this mean in terms of land use and new regulations?

Possible actions:
_____
_____

Issue 9 Product

Think about your products. What is it you are actually producing? Good design can reduce environmental impacts from the start of a product's life to its final disposal or reuse:

9.1 What are your main products?

9.2 Where are the main environmental impacts through its life cycle?

9.3 Are products designed to minimize environmental impacts and reduce energy and waste?

9.4 Is environment considered in product design and development?

9.5 Are product design issues discussed with customers/suppliers?

9.6 What management procedures are in place to identify improvements in this area?

Possible actions:

_____

_____

_____

Narratives

Smallco

Issue 10 Process

The way you operate your processes will impact on the environment. (Figure 4.2 in the next chapter shows a simple organizational process model.) Sketch out your own model from the ideas here and add environmental impacts:

10.1 What are the main processes you operate?

10.2 Where is the main environmental impact(s) – actual or potential? Past, present and future?

10.3 Are production processes operated to minimize environmental impacts and reduce energy and waste?

10.4 Are operators always trained in best practices?

10.5 Is environment considered in process design and development?

10.6 Are design issues discussed with customers / suppliers?

10.7 What management procedures are in place to identify improvements in this area?

Possible actions:

_____

Issue 11 Raw materials

The nature of your business will dictate what you buy in as raw material for the production process. A listing of raw materials item by item (including a break down of sub assemblies, components and materials used by service providers) will give you an inventory to work from. Think about what comes in and what goes out. You should be able to find out systematically the source of your raw materials and begin to judge the environmental impacts caused along the way. There might be reasonable alternatives that your business could use to reduce these impacts:

11.1 What raw materials are purchased and where do they come from?

11.2 Do suppliers have an environmental policy?

11.3 How efficiently are raw materials used in production processes?

11.4 What is the purchasing policy with respect to raw materials?

11.5 Are materials minimization plans in place?

11.6 Are there options to reduce environmental impacts of raw materials?

Possible actions:

Issue 12 Hazardous materials

A quick tour will reveal just what hazardous materials you are using and how much of them you (or subcontractors) use in your business. Although you are probably already dealing with the health and safety issues associated with their use, there will be environmental issues associated with them which will need to be managed in your EMS:

12.1 What hazardous materials do you use?

12.2 Where do you use them?

12.3 How much do you use?

12.4 Are all staff who need to be trained actually trained?

12.5 Assuming you satisfy all relevant health and safety considerations, what are the environmental implications of the use of these substances?

12.6 Do you have any choices to reduce their use, for example, by switching to safer alternatives? Can the suppliers of these substances help here?

Possible actions:

Things to think about

Issue 13 Water

Water is a major service to the business and an often overlooked resource. You just turn on a tap and it's there! It is easy to forget that you usually pay for it twice; once when it comes into the business and is used for whatever purpose and once again when you discharge it into the sewers or have it taken away by a waste disposal contractor:

13.1   How much water is used and what is the cost?
13.2   Where is it used?
13.3   How is it used?
13.4   How is it discharged and how much does this cost?
13.5   What regulatory consents and constraints are in place?
13.6   What could be changed to reduce water usage?
13.7   Could water be reused?

Possible actions:

_____

_____

_____

Issue 14 Energy and fuels

Energy, which keeps the business running, is a strange and rather abstract concept. The bills for it are not. In many ways, energy is simply a label for our own convenience and a measure of something else happening; for example, the use of electricity in kilowatts or the calories (heat) in various items of food. Organizations can usually cut 10 per cent off the energy bill without capital expenditure, another 10 per cent from investment which rapidly pays for itself and another 10 per cent by investing in the longer term. Saving money and reducing pollution can go hand in hand. Remember, every £1 saved by efficiency measures is a bottom line saving.

14.1   How much energy is used and what are the unit costs of:
        (a)   electricity?
        (b)   gas?
        (c)   oil?
14.2   What is it used for?
14.3   What are the demand variations?
14.4   What management controls are in place?
14.5   What relationships exist with energy suppliers/advice on energy conservation?
14.6   Does equipment purchasing and maintenance include energy issues?
14.7   What opportunities exist for energy savings?
        (a)   minimization?
        (b)   energy recovery?

(c)   energy reuse?

(d)   renewable energy sources?

Possible actions:

_____

_____

_____

Issue 15 Stakeholders

In Chapter 2 we considered the drivers for environmental management. These drivers included both internal and external pressures which need to be evaluated in the establishment of an EMS. Several of these stakeholder needs are addressed by the questions above. This final 'issue area' mops up the remaining key stakeholders (or 'interested parties' as ISO 14001 calls them) so that their needs can be considered as part of the initial environmental review:

15.1   What are the needs of the customers themselves? What do they expect and what might they expect in the future?

15.2   What are the demands and expectations of your banks and insurance companies?

15.3   People care about the environment. Have you identified all your stakeholders and their concerns? How are relationships with 'the stakeholders' managed?

15.4   What measures do you take to involve staff and (where appropriate) trade unions in the process of environmental management? What are their expectations in these areas and how will you deal with them?

Possible actions:

_____

_____

_____

What level of analysis is required?

To devise an effective EMS for your organization you need environmental knowledge and confidence. In the IER, by taking stock you have taken the first step on the road to success and begun the planning process to involve others in the work. Once you have taken stock, the next step is to find out what the information means in the context of your organization and what the implications are of your past, present and future environmental decisions or indecisions! Making organizational sense of the information will enable you to tackle the issues with new and growing confidence. Remember too that the reward for solving some problems is not a quiet life – it's a better class of problem!

You should now have a wealth of up to date and accurate information which reflects the way your organization manages the environment and the environmental issues associated with its business operations. Now is the time to put it into context in ways which make sense for the business. You should be able to make distinctions between what you must do – such as comply with all relevant legislation – and what you can do – such as invest in new equipment – to make improvements to your management of environmental aspects of your business. Some basic guidelines include:

- Whatever you do with the information you have collected, and by now there will be a lot of it, keep the interpretation as simple as you can.
- Think about the information in terms of people, systems, technology and your site. For each of these think about:
  - Legislation and standards. Do we meet them, exceed them or fall short?
  - Is there any established best practice? What is it? How do we measure up? This is usually referred to as benchmarking.
  - What do you need to do, what can you actually do and how much might it all cost?
  - Which procedures might need adjusting, completely changing or even establishing from scratch?
  - How are you actually performing?
  - Draw some key conclusions from the work, set these out and engage colleagues in discussions about the findings.

Later, in Chapter 4, we will extend the consideration of the issues we have examined by identifying the related environmental aspects and potential or actual impacts. It is here, by managing significant environmental aspects of the business and setting realistic objectives and targets to handle them, that the EMS really begins to mature.

 Narratives

Smallco

Despite the Ops Manager's efforts to communicate with management colleagues throughout his five step approach noted in Chapter 2 *(1. To further investigate phased implementation; 2. To map out an implementation plan; 3. To begin discussions; 4. To gather cost savings ideas; and 5. To present skeletal plan)*, there is still much confusion and disagreement about what to do next. Some managers still perceive an EMS to be a waste of time while others agree in principle but say they have no time in practice. An IER focusing on developing simple environmental indicators to start building some hard evidence about what is going on in Smallco and what could be achieved is the logical next step. Scoping the IER so it makes sense for Smallco is of paramount importance. While the Ops Manager is clear about what needs to be done there is an important issue to ensure cooperation from colleagues. The Ops Manager decides to adopt a two stage plan to scope the EMS.

Stage 1. At the next management meeting he does two things:

1   Presents back the findings of his five tasks noted above. His presentation emphasizes opportunities and realities – for example the benefits of a phased implementation (whether supported by an external scheme or not) and the need to maintain legal compliance – deliberately skims over the skeletal plan he has worked on and finishes with some hard hitting examples of the benefits of best practice gleaned from similar sized businesses. Discussions with the customer company have gone well and help has been promised on legal compliance and environmental training.

2   Sets each manager some tasks for the next monthly meeting. In reality these are a first step IER in the form of a straightforward and focused questionnaire for each manager. These are based on the ideas in Chapter 3. The Ops Manager also offers to help in their completion and defines exactly what help means so they are not simply handed back! Thankfully, the MD supports this approach and requests that the work is done a week before the next meeting so that the Ops Manager can collate the data.

Stage 2. At the next but one meeting the Ops Manager presents back the information, making sure that the source of actual work done to provide it and the ideas contained in it are acknowledged. Much discussion ensues, not so much on what needs to be done but on how and when it can be done. The outcome of the meeting is pretty close to what is needed to scope Smallco's IER.

The Ops Manager spends an hour or so polishing up his notes to provide an IER plan with pertinent EPIs (including some suggested areas which might result in quick wins for Smallco) which is circulated with the management meeting minutes, which are of course endorsed by the MD.

BIG Inc

The project manager already had some intimation that scoping the review for a service oriented company would not be easy; the environmental impacts would be largely indirect (see Chapter 4) and associated with the design influence that the company could bring to bear on their client's installations. Bearing in mind that the directors had decided that they would want to preserve the option of applying for EMAS at some future point, the project manager decided to incorporate the EMAS criteria in Annex I.C of the regulation, and to keep the IER scope as wide as possible to cover the entire organization, including the European offices and the home-workers.

Table 3.3 Smallco draft phased implementation of EMS (after ACORN and BS 8555)

Phase 1 **Commitment & establishing our baseline**	1 Management commitment
	2 Baseline assessment – A basic IER plus EPE
	3 Develop Smallco environmental policy
	4 Develop simple environmental indicators
	5 Develop an initial draft EMS implementation plan
	6 Competence, training and awareness needs
	7 Continual improvement – what makes best sense?
Audit to determine progress...	
Phase 2 **Compliance with legal & other requirements**	1 Catalogue relevant legal requirements
	2 Research any 'other' requirements
	3 Compliance check(s)
	4 Develop appropriate compliance indicators
Audit to determine progress...	
Phase 3 **Developing objectives, targets and programmes**	1 Evaluation of Smallco aspects and impacts
	2 Finalize and sign off environmental policy
	3 Develop appropriate objectives and targets
	4 Set up environmental performance indicators
	5 Develop our environmental management programmes
	6 Develop our operational control procedures
	7 Launch policy internally; objectives and targets and indicators
Audit to determine progress...	
Phase 4 **Implementation & operation of the EMS**	1 Define management structure and responsibility for the EMS
	2 Implement identified training, awareness and competence programme

SETTING UP

Table 3.3 (*continued*)

	3 Implement agreed communications – internal and external
	4 Set up document and record systems
	5 Implement and test emergency procedures
	6 Utilize and further develop indicators as necessary
Audit to determine progress...	
Phase 5 **Checking, audit and review**	1 Develop EMS audit programme
	2 Carry out corrective and preventive actions as required
	3 Set management review meeting(s) and agenda
	4 Suggest most beneficial ways to improve performance
	5 Use EPE work to suggest improving our EMS and review objectives and targets as necessary
Audit to determine progress...	
Phase 6 **EMAS certification**	Consider rationale for optional external assessment of our EMS for BS EN ISO 14001: 2004
	To be discussed with senior management

Questionnaires issued to the managers at the European offices were inconclusive, as they felt that they had little direct control over their environmental impacts, especially as even the facilities and services were supplied by a separate company. Those issued to home-workers did not reveal a particularly clear pattern of impacts, the answers being largely subjective. Though there was no problem identifying external stakeholders, the feedback from them in the form of questionnaires and interviews was not helpful because they didn't define their expectations particularly clearly. Many were surprised to be receiving anything from the company in the first place, as they hadn't considered that the consultancy had any great environmental effects, or that a review was entirely relevant.

Trade secrets

- Plan to spend at least 40 per cent of your total project time on the IER. This investment of time will pay off and the EMS will be firmly based on the reality for your organization.

- Always carry out an IER before you draft your environmental policy. The policy is a public domain document and stakeholders are likely to pick you up on any fuzzy or misleading statements. Once again, the IER provides the underpinning for a meaningful policy document.
- Get people involved and spread the workload. The IER stage of installing an EMS will provide lots of opportunities to engage colleagues and perhaps other stakeholders. Don't miss the opportunity for this early engagement. It will make EMS implementation much easier later on.

Things to think about

- Guidance documents in the ISO 14000 series are well worth looking at, particularly *ISO 14004: 2004 Environmental management systems – General guidelines on principles, systems and supporting techniques*. This provides a wealth of information on IER.
- A rigorous IER and ongoing re-evaluation of your environmental issues will give you an excellent foundation to build on if you plan to report externally on your environmental performance at some future point.
- Phased implementation, as suggested by BS 8555, may make sound business sense for smaller businesses but this might not be clear until you have completed your IER and everyone appreciates the nature of the task.
- Expect to see a rise in the number of organizations that don't just have one IER (despite the name) – they'll plan another on a three to five year cycle. Why so? Companies and management move on. A new generation of managers (how many of your organization's senior management team are still there after five years?) will have a different perspective and the organizations themselves will also have changed. The technique revisits old assumptions, does away with wish lists and injects hardcore reality into flagging EMSs.

Planning

4

Aspects and Impacts: What Are We Trying to Manage?

This chapter is closely linked to the previous work of taking stock of the organization's environmental position. This work will have established what is going on from an environmental perspective. For a new EMS this provides the foundation to address exactly what you are going to manage. For a more mature system there may be changes to activities, products and services or scope of the EMS. In either case the starting point is to establish the organization's environmental aspects and impacts for the defined scope of the EMS and to focus management effort on those aspects that are classified as being significant. This chapter provides the more detailed analysis required.

☑ ISO/EMAS quick check

Area of EMS	ISO 14001	EMAS
Environmental aspects	4.3.1	Annex I-A.3.1
Objectives, targets and programme(s)	4.3.3	Annex I-A.3.3 & A.3.4
Competence, training and awareness	4.4.2	Annex I-A.4.2
Communication	4.4.3	Annex I-A.4.3
Operational control	4.4.6	Annex I-A.4.6
Monitoring and measurement	4.5.1	Annex I-A.5.1

▢ Chapter executive summary

Move to the next chapter when you can answer all of the following questions in ways which make sense to you and your organization:

Is there a systematic way to identify aspects and impacts?
What do they mean for the organization?
What will effective prioritization look like?
How do they link with the EMS?

You are here

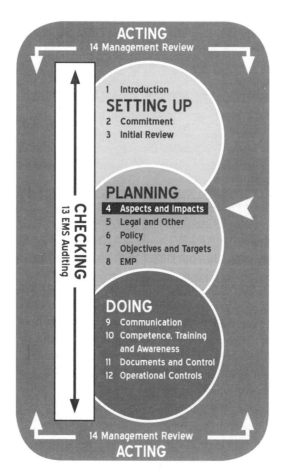

Toolkit requirements

To work through this chapter you will need the following knowledge and skills:

Knowledge

- what your organization does and the scope of your EMS;
- the key environmental issues for your business;
- broad environmental issues.

Skills

- information analysis;
- prioritizing;
- management judgement.

Context: What are aspects and impacts?

If you don't know where you are going, how will you know you've got there? In this section we will introduce and develop the knowledge and understanding to answer a fundamental question for anyone setting up an EMS: Just what exactly are we trying to manage? The simple response would be 'Everything!' requiring sufficient resources at your disposal to allow work to take place on all fronts simultaneously. In the real world this is unlikely. In building an EMS this siege approach is not particularly appropriate. A more planned approach to management will help deliver a more effective EMS.

The 'ISO/EMAS Quick check' above outlines the importance of aspects and impacts analysis. The outcome of this work is essential to establish the rest of the EMS and make adjustments as the EMS evolves. It is important to understand the relationship between aspect and impact:

Table 4.1 EMS: Aspects and impacts meanings

Aspect: The cause

Element of an organization's activities, products or services that can interact with the environment. For example, using energy means burning fossil fuels and ultimately. . .

Impact: The effect

Any change to the environment, whether adverse or beneficial, wholly or partially resulting from an organizations activities, products or services. In the example above the impact of the burning of fossil fuels will be resource depletion, effects on local air quality and a contribution to global warming. 'Cause and Effect' provides a useful overview of these issues which embraces both environmental management standards. What are the aspects of the business that actually or potentially result in an impact on the environment?

Getting to grips with cause/aspect and effect/impact can best be demonstrated by some upside down thinking. By starting with an effect or impact on the environment we can track back to the root cause and identify what exactly is the basis for the effect.

The chart below provides several examples:

Table 4.2 Root cause of environmental effects

Environmental effect impact	The root cause (ie the 'aspect' of an organization's activities, products or services that could be causing the effect)
Global warming	The use of fossil fuels, the burning of which releases carbon dioxide. This could be, for example, in heating systems; electricity use (someone else does the burning at the power station); transport fuel and so on. . .
Ground water contamination	Pollution from toxic releases; increasing the biological or chemical oxygen demand through discharges; increasing the temperature of watercourses
Species loss/habitat destruction	Development - new buildings, civil engineering, etc
Air quality	Processes that involve incineration; release of volatile compounds; release of toxic compounds
Noise pollution	The operation of plant and equipment; transport issues and so on...
Contaminated land	Past historical activities on the site which may have been 'inherited' when the site was purchased

Clearly this table is incomplete and there are many more categories of environmental impacts. Think about how your organization's activities might impact on air, water, land, biodiversity, visual intrusion, noise and so on. It is worthwhile considering your own situation and brainstorming the environmental impacts that your organization may be capable of producing in normal, abnormal and emergency situations and then analysing just what aspects of your activities, products or services are or could be the root cause of these possible impacts.

Whichever management standard you might eventually subscribe to, whichever phased implementation scheme route you take, or even if you decide to self-declare, there are some common methodologies to identify environmental aspects and impacts that are extremely useful. These are set out in the next section.

Tasks

Is there a systematic way to identify aspects and impacts?

Your work on the baseline environmental review – described in the previous chapter – will have identified what is going on in your organization from an environmental perspective. This work on environmental aspects and impacts is a logical extension of the review and begins the process of answering the fundamental question raised in the title of this chapter: What are we trying to manage? This first step is simply about identifying aspects and impacts. The interpretation of what they mean for the organization and what to do about them in the context of the EMS comes later. In reality, everything that goes on in an organization – from people washing their hands to sourcing raw materials and the actual 'manufacturing' of a product or the delivery of a service – will have some form of impact on the environment.

Data	→	Processes	→	Evaluation
Gather the relevant data about your organization		Understand what goes on in your organization		Identify significant impacts

What is important in terms of the EMS is to ensure that the approach taken to identify aspects and impacts – the causes and environmental effects of what you do in your business – is rigorous and consistent throughout the organization. Three tools and techniques help provide guidance:

- organizational analysis;
- life cycle assessment (LCA);
- issue by issue.

Organizational analysis

Start by constructing the big picture of your organization. Most organizations can be simplified into three distinct steps:

Step 1 They buy in a range of 'raw' materials from their suppliers.
Step 2 They produce something with them using a range of techniques and processes.
Step 3 They sell goods or services to their customers.

A wide range of staff with diverse and complementary skills is involved in each stage of the three step business. By working in concert, the outcome is a healthy and economically sustainable business. We'll come on to the training and communication issues that might be needed later in the book.

An effective way to carry out your organizational analysis is to take three sheets of flip-chart paper and a stack of Post-it™ notes. On each flip-chart sheet write either 'BUY', 'PRODUCE' or 'SELL'; pin them up in this order. Now explore each in turn. Use a Post-it™ note every time you consider an issue and stick it up on the relevant flip-chart sheet. This structured brainstorm works well as a team approach and will help to ensure a range of perspectives and that all issues are covered. As the charts fill up take time to move the Post-it™ notes around and group related issues together. Use a flip-chart pen to draw in the links between individual issues or issue groupings. By the end of the exercise you will have constructed a text-based representation of what is going on in the organization or what has changed since you last reviewed activities from an environmental perspective. This is represented diagrammatically in Figure 4.1.

The concluding part of the analysis is to consider the links between what is going on in the organization and what this means in terms of the actual or potential environmental impact. Again, at this stage it is simply a case of establishing what the environmental impacts might actually be. Don't be tempted to interpret what they mean for the organization and try to find solutions just yet. This technique can also be used very effectively for one component of a whole organization, for example, an individual operating unit, process or office.

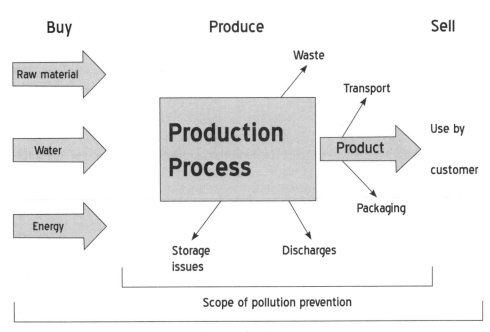

Figure 4.1 Organizational analysis

The second technique – life cycle assessment (LCA) – provides the link between what is going on in the organization with the potential environmental impacts of this BUY, PRODUCE, SELL activity.

Life cycle assessment

Quite simply this means thinking about a product or operation from its cradle (where it is born) to its grave (where it dies). In many ways it is an extension of the organizational analysis step just described, which factors in the linkages with the environment and so identifies potential and actual environmental impacts of organizational activity; present, past or planned.

The idea of life cycle analysis can be illustrated by considering a disposable ball-point pen. Many raw materials are used in its manufacture, for example, those derived from petrochemicals and there is an energy requirement to fuel the manufacturing process. For each input to the pen there is an environmental link. The extraction of oil and its eventual reprocessing into plastic feedstock has an environmental impact. More fossil fuel is used to generate electrical energy to drive the production process and yet more fuel for transportation of the pens to market. There are environmental impacts here too which can be described in terms of resource depletion, greenhouse contribution and perhaps local air quality. Finally the pen is consigned to the bin and probably a landfill site where its environmental impact (a small but cumulative one) is in terms of waste burden.

The pen example is illustrated in Table 4.3. By taking each activity, product or service identified by the organizational analysis and considering their life cycles, you can unveil their actual or potential environmental impacts. It is important to consider the actual or potential environmental impacts at all stages both in the life cycle of a product or service and in all facets of an organization's. This allows for more detailed 'What happens if?' scenario planning and will ensure a thorough evaluation to provide the intelligence for management decisions. A simple risk analysis technique is discussed later in this chapter (p66).

The analysis of impacts, aspects and significance will need to consider and evaluate not only normal day-to-day operating conditions, but also abnormal and emergency situations. For example, abnormal conditions could include rush jobs, machine or production line failures, maintenance operations or waste transfer operations outside the normal operating scope of the plant. Emergency situations might include accidents such as a fire, a breach of a storage tank, a yard spill or a major operator error. One way to evaluate the difference between 'normal' and 'abnormal' operating conditions is to use Table 4.4. For any activity consider its planning, the operating criteria you have established and the output from the process or activity. If any one of the three areas is 'uncontrolled' then you have an abnormal condition. If any two are uncontrolled then an emergency situation is likely to ensue rapidly. Many thanks are due to John Moulding-Dyas of Balfour Beatty for letting us reproduce this useful table.

If the management system is to be effective, then planning for abnormal or emergency situations must take place and judgements made on dealing with the aspects of these situations which might cause a significant environmental impact.

Table 4.3 Life cycle assessment (disposable ballpoint pen)

Activity	Environmental aspect	Environmental impact
Obtaining the raw materials	Extraction of oil	Natural resource depletion Potential water contamination Ecological habitat damage
Obtaining the raw materials	Extraction of metals	Natural resource depletion Potential water contamination Ecological habitat damage Visual intrusion
Manufacturing the plastic feedstock	Energy consumption	Greenhouse effect Local air quality Noise
Manufacturing the plastic feedstock	Use of chemicals, solvents, etc	Potential air and water contamination Local air quality
Manufacturing the pen	Energy consumption Paper and card	Greenhouse effect Local air quality Noise
Packaging the pen(s)	Energy consumption Paper and card	Greenhouse effect Local air quality Ecological habitat damage
Transporting the pens	Energy consumption	Greenhouse effect Local air quality Noise
Disposal of the pens	Landfill	Ecological habitat damage Visual intrusion Waste burden

Table 4.4 Evaluating normal and abnormal operating conditions

Planning	Operating criteria	Output	Condition
Controlled	Controlled	Controlled	Normal
Uncontrolled	Controlled	Controlled	Abnormal
Controlled	Uncontrolled	Controlled	Abnormal
Controlled	Controlled	Uncontrolled	Abnormal
Any two 'uncontrolled'			Emergency

PLANNING

A final note on this extended thinking is to consider that a business is not a static organism but is constantly evolving and changing the nature of its operations. Your analysis of aspects and impacts should take account not only of the existing realities of operations and activities but also what has gone before and what might follow in the future. For example, the site might well have a past history of use by yourself or others in ways which might mean a legacy of pollution or land contamination. Future plans might include expansion or selling a site, new products and services, new plant and new operations; all of which will potentially have new environmental aspects and impacts. Many companies count the sale value of their land as one of their assets; a contamination problem would seriously undermine the realizable assets of any organization. Your evaluation must consider these variants which are a reality for any business and are fundamental areas for attention to ensure a rigorous analysis of environmental aspects and impacts.

The third technique, mapping what is going on against the organization's key environmental issues, provides the next step in the analysis process.

Issue by issue

In Chapter 3, The initial environmental review, we examined 15 key environmental issues, ranging from management issues to stakeholders and considered these in relation to the organization. The next step in the identification of environmental aspects and impacts is to map these key issues against the organizational analysis. This will complete the picture of the organization's environmental aspects and impacts and so enable you to assess which are significant and will thus need to be embraced by the EMS.

Taking each issue in turn and considering the organization's aspects and impacts which relate to it (derived from the previous two activities) will enable a detailed identification and recategorization of aspects and impacts. The organizational pattern imposed by considering aspects and impacts issue by issue will provide a practical framework for the EMS. The 15 issue areas help simplify thinking and provide a collective focus for presentation and discussion with colleagues who will later be explicitly involved in the delivery and management of the new EMS. Before moving on, extend the table below to develop this third technique for those issues you analysed in your IER in Chapter 3.

Issue: see Chapter 2 for the key issues	*Aspect* of your business that results in this issue being relevant	*Impact* on the environment: what effect does this aspect have?	How might you manage the aspect of your business to reduce the impact?

Having identified it, can you control it? If you can't control it, can you influence it? Does anyone expect you to? These are the next questions you must ask yourself about each issue. It now makes sense to consider those areas a business has control and influence over. The outcome of the life cycle analysis of the activities, products and services of the business might have suggested numerous direct and indirect global links, all of which have some actual or potential impact. A decision has to be made about what is reasonable for the business to deal with in its EMS and what is outside its boundary of control. Considering 'control and influence' can not only help you to prioritize (see below) but is also a useful approach to deal with this question of defining the scope of the EMS. Your boundary of control is likely to form the core of your EMS and be amplified by considering your boundary of influence.

In Figure 4.2 there are issues which are under the direct control of the organization, for example, how it actually operates its production processes. For other issues the organization might have some indirect control and influence, for example, in the sourcing of raw materials from suppliers who have considered their own environmental position or in how its goods and services are actually used by the consumer. Drawing the boundaries is an important task in the scoping of an EMS. You must satisfy yourself and, ultimately, perhaps an external auditor that the process you have used is watertight, remembering that, as well as suppliers and customers changing, knowledge and understanding of the issues does too.

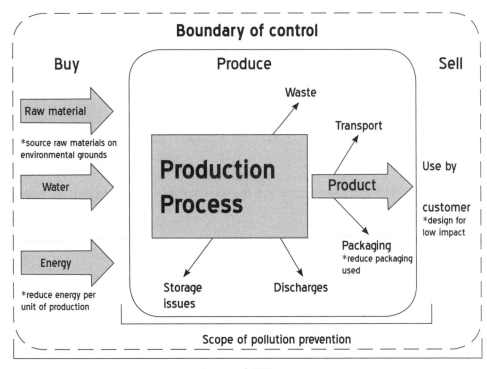

Figure 4.2 Boundaries of control

What do they mean for the organization?

In reality, everything you do – from driving to work to 'manufacturing' a product or service – will have an impact on the environment. Life couldn't go on without making an impact. For the business establishing an EMS, making sure that resources are channelled to those areas that need most attention is paramount. It is highly unlikely that all impacts can be managed simultaneously. It is therefore important both to know what your impacts are – as established by the first part of this section – and to consider those that are significant in the context of your business. The opportunity to do something about them is provided by your understanding of the aspects of your operations that produce or potentially produce these impacts in the first place. Remember, you are seeking to manage the aspect and not the impact. The purpose of the EMS is to work at managing the environmental aspects of your business effectively.

An analysis of findings can be carried out in two steps. Step one provides a quick overview and will help to reveal the major areas requiring attention. Step two allows a more detailed evaluation and in reality may initially be applied to only those major impacts revealed by step one. This also reflects the approaches most commonly used in phased EMS implementation schemes.

Step 1 A first assessment of the potential environmental impacts

A broad brushstrokes approach to analysing your impacts can be carried out by reflecting on three evaluation criteria:

- *Size* of the potential problem: how much damage could be done? For example, a spill of oil on to an unbunded yard next to a trout stream could potentially become a very significant impact on that ecosystem. An operative flushing paint thinners down the toilet, although having some impact, might be a one off and hopefully not a regular procedure!
- *Likelihood* of a problem occurring: how possible is it? You could, for example, know that operatives are well trained in emergency procedures and would carry out the necessary actions almost as a reflex.
- *Hazard*: how hazardous is the potential impact? Some potential impacts will be more hazardous than others. A spill of organic solvents may not only damage the local ecosystem, it may also threaten workforce health and drinking water quality too.

Take each identified environmental impact in turn and reflect on each of these evaluation criteria. A simple scoring system can be utilized to classify the impacts. Use a simple 1, 2, 3 scoring system where 1 means the issue is of no concern and 3 means an environmental impact would cause major distress to the environment, the business or external parties. Now simply multiply the three evaluation criteria scores together to obtain a total score for each issue under consideration:

$$S \times L \times H = \text{issue score}$$

The maximum score is therefore 27 and the minimum score is 1. This first step in evaluating environmental impacts will begin to reveal those that will need to be addressed by the EMS.

Step 2 A more detailed evaluation

This more detailed evaluation uses more criteria to extend the analysis. For each identified environmental impact we can consider a number of factors to form a judgement on its significance. These fall into three main groups: those that relate to the environment itself; those that are internal business concerns; and finally those that are external business concerns. By running each identified environmental impact through Table 4.5 and applying a simple scoring system, impacts can be ranked and so an informed judgement can be made on which are significant.

Having run each impact, whether actual or potential, through the table and adding up the scores you should now be in possession of a ranked listing of your environmental impacts. This list is a major building block for the developing EMS. It becomes the agenda for establishing objectives, targets and programmes – dealt with in Chapters 7 and 8 – which will bring the management system to life and demonstrate a proactive approach to business-focused environmental management. This is exactly what is embodied by ISO 14001 and EMAS.

What will effective prioritization look like?

Finally, prioritization will help by sorting out the areas which must be addressed immediately by the EMS to provide direct guidance on setting meaningful objectives and targets.

Any review of the organization's aspects and impacts will reveal many areas for attention. Even with immense enthusiasm and infinite resources it may still not be possible for a business to do everything at once. It is highly likely that such a 'blunderbuss' approach would not make systematic management sense either. Prioritizing for environmental management can follow a simple three step process: 'Must do', 'Will do' and 'Could do'. We can begin to define what might be meant by each of the three classifications, for example:

- **'Must dos'**: these actions could be legal requirements, cost saving necessities, issues which are significant environmental impacts or customer demands. (How do you know they are 'Must dos'? See Chapter 5)
- **'Will dos'**: these actions could be a short-term investment which would minimize pollution, bringing immediate cost savings.
- **'Could dos'**: these actions could be longer term management programmes which have been identified as effective investments and will bring measurable benefits in both fiscal and environmental terms.

The goal of this 'Must do', 'Could do', 'Will do' prioritizing is really about prioritizing the priorities! What is needed is a simple system to set key priorities. In the activity above – issue by issue, step two – we evaluated several key factors under the headings of 'Environment', 'Internal business concerns' and 'External business concerns'. This evaluation, taken issue

Table 4.5 Ranking environmental impacts

Factors to consider	Score (*)
Environmental factors	
What do we know? Advice from regulators, industry associations and scientific knowledge may point to an issue in your organization with potential to cause an environmental problem or impact. For example, use/disposal of an organic solvent will have a high ranking	
What is the receiving environment? Consider the sensitivity of the local environment which may be the recipient of your impacts. A site of ecological importance on your doorstep might mean a high score here	
What is your contribution? Think locally, regionally and globally. Just what contribution are you making in relation to these different levels?	
If something were to happen what is the potential size and period? Consider how big and how long an impact you might create would be. What would this do to the external environment? Could any impact be easily undone?	
Is it likely to happen? This is the final 'environmental' consideration. Something you do may be likely to cause a small environmental impact on a regular basis or perhaps a major environmental impact in an emergency situation	
Internal business concerns	
What are the environmental regulatory requirements? You will be bound by various legal requirements. What do these mean if you create an impact?	
What health and safety issues apply? Formal and informal requirements will apply. What happens if these are breached?	
What happens to production and what will it cost if an impact occurs? If production has to stop to ameliorate the impact, what effect will this have on production and for how long? How much will this cost, short and long term?	
What is industry best practice? The business may be a signatory to a best practice scheme. An unmanaged impact may undermine or destroy credibility	
How will staff react? What will happen to staff morale and commitment if an impact occurs?	
External business concerns	
What will the bank or insurers say? Their expectations will be for at least compliance with the law. What will they dictate if things go wrong?	
What will the public say? The public expect higher and higher environmental standards from business. How might they react to an environmental impact?	
What will customers say? Customers may already be demanding environmental excellence from you. How might they react to bad practice?	
Total score:	

(*) Ths scoring system: Use the same system as before where 1 means the issue is of no concern at all and 3 signifies that an environmental impact would cause major distress to the environment, the business or external parties.

by issue, will have helped rank environmental impacts and allowed a judgement to be made about which are significant and require management interventions to ensure an effective EMS is created. This final step in the process of managing environmental impacts provides the management focus from which realistic objectives and targets can be established as the EMS is built.

For the first part of this prioritizing of areas requiring attention, re-examine those impacts which have been uncovered as significant in the earlier analysis. The environmental impact is, remember, the effect and its cause is the environmental aspect of the business activity. For example, 'Waste Burden' is the effect or *impact* on the environment and its cause or *aspect* is the use of natural resources by the business. The efficacy with which the business operates will determine just how much waste burden it creates. Table 4.6 provides a useful way to prioritize those significant impacts revealed in the earlier analysis. It makes good sense to make the links with those aspects of the business activity that are the root cause of the identified impacts for it is *only* at the aspects level that the organization can make a real difference to its impacts.

In many ways this final step is your own extended version of where we started in this chapter. Clearly, a complete version of the table below is really just the executive summary of a substantive body of investigative information. The small print is really in your domain to complete the picture and so provide the foundation from which realistic objectives and targets can be set using the guidance in the next section. By way of illustration two examples are given below based on real businesses who have designed their EMS to address and manage their significant environmental aspects and impacts.

How do they link with the EMS?

Making sense of the aspects of your operations that produce or potentially produce environmental impacts is the next step in the process. In the last activity we made the links between the environmental impact and the environmental aspect. This linking of the two is essential because any management intervention must seek to reduce or eliminate significant environmental impacts of the business in ways that are relevant to core business activities. Realistically, this can only be achieved by managing those aspects which cause the significant impacts: known as the significant environmental aspects. An effective EMS must deal with managing these significant aspects of the business. It is here that the business can make a difference. If this is to be done well then a systematic approach will ensure that the energy and effort needed to adjust existing management systems or perhaps create new ones results in outcomes which make business as well as environmental sense.

Both ISO 14001 and EMAS require that the management system deals with environmental aspects and impacts of the business as shown in Table 4.7. The systematic approach to identifying aspects and impacts will ensure that the implementation of the EMS, through setting objectives and targets and the management programmes to bring these to life, will bring both business benefits and environmental improvements through reduced environmental impacts. One way to explore further the business activities that cause the effect (ie the environmental aspects) is to sketch out an 'ishikawa' or fish-bone chart. This is a structured brainstorm of all the possible causes or aspects of business activity which may

Table 4.6 Environmental impacts, aspects and key areas for attention

Significance Score From Table 4.5	Description of impact	Must do: these actions could be legal requirements, cost saving necessities or customer demands	Will do: these actions could be a short-term investment which would bring immediate cost savings	Could do: these actions could be longer term management programmes which have been identified as effective investments and will bring measurable benefits	Aspect: what is the business activity that is the cause of these impacts?	Options: what management options already exist or could exist?

...And so on for all significant environmental impacts

result in an environmental impact. One example is given in Figure 4.3. This technique is an effective way of identifying all the possible aspects of a particular environmental impact and highlighting areas for possible management actions.

Table 4.7 ISO 14001 and EMAS requirements

Area of EMS	ISO and EMAS clauses	
	ISO 14001 refs	EMAS refs
Policy – '. . . environmental impacts of activities. . .'	4.2(a)	I-A.2
Environmental aspects – procedures to identify significant impacts	4.3.1	I-A.3.1
Objectives and targets – set these based on significant aspects	4.3.3	I-A.3.3
Environmental Management programmes – the action plan to realize objectives and targets	4.3.3	I-A.3.4
Competence, training and awareness – procedures to make employees able to deal with them	4.4.2(b)	I-A.4.2
Communication – with regard to its environmental aspects	4.4.3	I-A.4.3
Operational control – procedures related to identified significant impacts	4.4.6	I-A.4.6
Monitoring and measurement – monitor and measure key characteristics of activities that can have a significant impact	4.5.1	I-A.5.1
Management review – addresses need for system changes and adjustments	4.6	I-A.6
All above relate to the activities, products and services of the organization		

There is also the important consideration in ISO 14001 of continual improvement discussed in Appendix IV. A thorough understanding of the aspects of an organization's activities will allow considered management choice in modifying these activities so that the associated environmental impacts are reduced or even eliminated. Across the whole organization this understanding of process will provide numerous opportunities for positive change and secure both business and environmental benefits.

 Narratives

Smallco

Careful planning using a phase implementation approach prevented Smallco from being overwhelmed by their analysis of environmental aspects and impacts. In total 37 issues emerged, some interlinked and some not. One of the senior managers joked that before

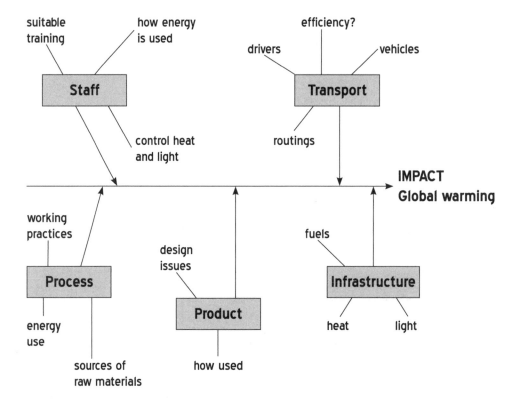

Figure 4.3 'Ishikawa' chart

the work he didn't realize they had anything 'environmental' going on at all! Despite the help from various managers to carry out the baseline assessment and list the environmental impacts and the business activities that actually or potentially caused them, the Ops Manager alone was left with the vital task of making sense of the data which he did by cutting the data in two ways.

1 He took a random sample of ten issues and did a quick calculation of how much it would cost to do anything and what the benefits might be.

2 Using Post-it™ notes he mapped out relationships between the 37 issues. Some clear patterns emerge particularly relating to discharges and noise. Delving deeper into the baseline IER work he found:

(a) Twice in the past eight months Smallco has contaminated surface water drains with discharges. Operator error – hosing down the yard – was blamed on both occasions, although this was due in part to poor storage. The local environmental health officer has indicated that Smallco must improve their management systems.

(b)   Local residents have complained about the increase in lorry movements, particularly early in the morning and late at night. As well as the environmental issue, noise, some parents are worried about children's safety going to and from the local school.

Armed with a re-drawn 'map' from his Post-it™ notes work and back-up data from the baseline assessment, the Ops Manager is ready to report to the senior management. The thrust of his report and recommendations is that there are three areas to work on and manage:

1   Those areas which need immediate attention and are causing environmental impacts and potentially put Smallco on the wrong side of the law.

2   Those areas which will be addressed in the next six months. The focus of attention will be on dealing with the root cause, so that limited resources are used to maximum benefit. A key message for staff and stakeholders is that Smallco seeks to become proactive in its environmental management work.

3   Consolidating the help offered by the supplier and making this best fit identified needs.

This work has completed the tasks in the first phase of a phased implementation. This approach, with its business focus, has been instrumental in overcoming the doubt expressed earlier by key colleagues in Smallco.

BIG Inc

Feedback from the review was piecemeal, though there had been some interesting developments in terms of identifying direct impacts. The most obvious one was energy consumption in one form or another, even though much of the energy management needed to be defined at the strategic level. Design aspects, transport and purchasing policies were quickly identified as the next most important areas. Overall, the project manager felt that the review said little and gave little direction. She was also picking up feedback from line managers that they felt the whole EMS project was largely meaningless for BIG Inc. This told her that unless she could dispel this belief, an EMS was highly unlikely to be supported, simply through ignorance.

As a response, she got permission and support from senior management to hold a half-day workshop for all managers (including representatives from European offices) as an add-on to the normal monthly business management meeting. In it she introduced an LCA model complete with an exercise in which teams of managers compiled their own LCA of the company and identified the environmental aspects of each of the operations in the cycle. The workshop was a great success, emphasizing to all present that, although most of BIG Inc's environmental impacts were indirect (ie not directly within their management control), the amount of indirect influence that the company had over environmental aspects in other parts of the supply chain was much greater than anyone had realized. The workshop produced several interesting ideas and suggestions from the managers, who now had no difficulty buying into the whole project.

Trade secrets

- Manage those issues that can and will make a measurable difference. Often these are not immediately obvious or glamorous; hence the importance of basing judgements on the IER and its analysis.
- Make absolutely sure those involved know how they can best contribute and keep them posted on how they are doing.
- Make strong links with what people do and can do. Help them to help you and the EMS. Make global links in ways which make local workplace sense.
- Consider how, and then talk to your stakeholders. They can often help in the short term with advice and guidance. In the medium term this approach will build productive relationships.
- Set your EMS up for success. People respond best to positive outcomes. Set achievable targets to deal with your environmental impacts. Clearly some must be achieved on short time-scales and might be challenging work. Others can be achieved by smaller incremental steps with milestones along the way.

Things to think about

- Strangely, there is no guidance in the ISO 14000 series of standards that relates specifically to aspects, impacts and their relationship. Expect other bodies (professional bodies, trade associations, etc) to fill the gap by creating and updating relevant material.
- Aspect and impact work may help to establish the primacy of environmental factors (over the social or even economic aspects) that relate to sustainability. The rise and rise of eco-footprinting as a headline technique bears this out.
- In the long term, environmental managers will either disappear by being 'mainstreamed' into business management as a whole or will become markedly more Establishment through the creation of a 'professional' infrastructure. Either way, it's good to remember that at the moment we still only have one planet to screw up.

5

Legislative and Regulatory Issues: What Must We Do?

An EMS can play a vital role in maintaining compliance with relevant legislation and regulation. If you are implementing an EMS you do not need to be a legal expert, but you will need to be able to uncover what legal requirements apply to you and your organization. Understanding how an EMS works in relationship with legislation, tracking the development of environmental regulations, and knowing where to look for further information are first steps towards ensuring that you stay inside the law and avoid the penalties. *Please note that this chapter is for general guidance only and that you will require the assistance of a legal professional to ensure that all your environmental liabilities have been identified.*

✓ ISO/EMAS quick check

Area of EMS	ISO 14001	EMAS
Environmental policy	4.2	Article 6 (Clause 4)
Legal and other requirements	4.3.2	Annex I-A.2
Objectives, targets and programme(s)	4.3.3	Annex I-A.3.2, Annex I-A.3.3
Competence, training and awareness	4.4.2	Annex I-A.3.4
Operational control	4.4.6	Annex I-A.4.2
Evaluation of compliance	4.5.2	Annex I-A.5.2
Non-conformity, corrective action and preventive action	4.5.3	Annex I-A.5.2
Control of records	4.5.4	Annex I-A.5.3

⬚ Chapter executive summary

Move to the next chapter when you can answer all of the following questions in ways which make sense to you and your organization:

> *Do you know how to identify relevant environmental legislation/regulation?*
> *Do you know how to structure your monitoring of legal developments?*
> *Do you know which elements of your EMS the law affects?*
> *Do you know what is meant by 'other requirements' in EMS standards?*
> *Do you know where to go for further information?*

☑ **You are here**

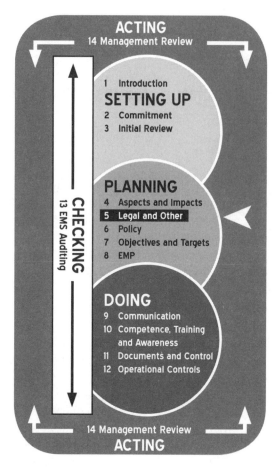

🔧 **Toolkit requirements**

To work through this chapter you will need the following knowledge and skills:

Knowledge

- the organization's environmental aspects of activities, products and services (past, present and future);
- policy commitments;
- views of interested parties;
- understanding of requirements of EMS standards (where applicable).

Skills

- knowledge/information management techniques (database handling/creation);
- information evaluation;
- accessing specialist knowledge (when required);
- procedure writing.

Context

One of the most often quoted reasons for development of a formal EMS has been repeated problems of maintaining company environmental performance inside the requirements of the law. It is important for any company to understand the exact details of which activities, products and services are affected by the environmental legislation. In addition, managers need to understand how such law develops and where to monitor such developments on a monthly, sometimes weekly, basis. Environmental legislation and its attendant regulations never stand still, but change in line with further scientific evidence, improving monitoring technology and greater general understanding.

Those companies that lose sight of legal developments can pay a heavy price, up to and including the demise of the organization if an apparently sudden and large investment in new plant is not accounted for in advance. Banks and finance houses look upon requests for money in such circumstances with a cold eye, regarding it not so much as the funding of environmental protection as the subsidizing of inadequate forward planning by the company management. As a result, it is useful for a manager to regard legislation in this area as much more than an obstacle to be overcome; a social expression of baseline acceptability, with a series of stringent penalties attached for under-achievers. It is better by far to see legislation, regulations and regulators as providing precise characteristics of the environmental aspects of the activities that need to be measured and subsequently managed.

The steps to acquire this knowledge are discussed below.

Inventory: Draw up a detailed list of activities, processes, substances, products and services that either (a) have a significant environmental aspect to them or (b) fall under the coverage of local, regional or national legislation.

Identify (relevance test): There may be some legislation that applies either under special circumstances only (not during normal working operations) or perhaps only to processes above a certain size or output. Simply because you have a boiler on site, does not necessarily mean that every single piece of air emission legislation actually applies. As a result, a relevance test may be extremely important in the early days of your management system. However, it is worth noting that exclusion clauses in regulations are relatively easy for a country's legislative system to change or remove altogether. Such changes can catch companies unaware unless carefully monitored. It should also be noted that not all environmental requirements are recorded in overtly 'environmental' legislation; it may also be part of health and safety, product or business law.

Characterize: Always match your organization's specific circumstances to the regulatory requirements, as there may be important information concerning monitoring and sampling requirements contained in licences, permits and consents issued by regulators. This will place a direct requirement on your ability to measure the impact you are seeking to manage. It is also important to compare your own forward planning and future projections with the trends in the appropriate legislative area.

Update: Understanding the relevant law is not a one-off process. Once you have established the link between your operations with the law that applies to them, you will still need to monitor legal development relatively closely. Don't restrict yourself to monitoring those elements identified in the IER (see Chapter 3), because new legislation, or changes to existing instruments can place requirements on processes and substances previously unaffected. Keep the focus of your monitoring wide enough to take in potential changes as well as those already identified.

If you intend to establish a formal EMS, identification of relevant legislation is a core element of such a system. It affects over eight different elements of ISO 14001, while EMAS goes even further, as companies can be de-registered by the appropriate national Competent Body for a breach of relevant regulatory requirements. It is also a key part of phased implementation schemes, occurring early in the implementation process for obvious reasons. Without this work having been carried out successfully, the remainder of the implementation could be seriously flawed.

 Tasks

Do you know how to identify relevant environmental legislation/regulation?

As no two sites and no two operational processes are the same, the precise catalogue of relevant environmental information that applies to your organization will be similarly varied. The most important goal to achieve is a relative match between the boundaries (or stated scope) of your EMS (see Chapter 4) and the law that applies to it. For instance, it would not be appropriate to design your EMS in such a way that it did not manage environmental impacts covered by legislation. As the law is to be treated as a performance baseline, it is vital to know exactly where that baseline lies. As a first survey of your operations, try answering the following questions, grouped for convenience under six headings. It is by no means a complete legal checklist, but when you can answer them all, it is likely that the scope of the legal issues affecting your site(s) will have been revealed enough for you to pursue matters in further depth.

1 Planning and land issues

- Are the boundaries of your EMS the same as your site physical boundaries? (If your stated EMS scope is bigger than the site boundaries the remainder of these questions will also apply to any other locations or activities covered by your EMS. If your EMS is smaller than the site boundaries, be clear who manages or has responsibility for the 'excluded' parts of the site and how you ensure adequate management of legal issues that may still affect your organization.)
- Is your organization the owner of the site, or the leasehold tenant?
- Do you have an up-to-date site plan (preferably showing information on below ground installations , drainage and storage facilities – eg oil tanks)?
- Do you know the details of your existing land use/planning permit and what it allows you to do?

- Are you storing hazardous materials anywhere on your site?
- What is your planning time horizon? Do you know what you are going to be doing on your site for the next five or ten years?
- Are you planning any changes of use of any part of your site?
- Are you intending to extend or alter any of the physical plant on the site?
- If you have a new project planned, have you carried out any assessment of the associated environmental impacts?
- Do you know what plans the local planning and development (zoning) authority have for the area in the vicinity of your site (ie within a 25-mile radius)?
- Do you know a planning (zoning) officer from your local planning authority by name?
- Have you spoken with them recently about land use/planning issues?
- Are you in a residential area, or is there residential development planned nearby?
- How complete is your understanding of the history of the site?
- Do you have the results of any previous land/ground water surveys?
- Do you know if you have any contaminated land/ground water on your site?
- Does your local planning authority know?
- If you have, have you (or previous land users) taken any steps to clean it up?
- Do you have any other land-related features of note, on or about the site (trees/rivers/underground streams/wild habitats/ecological sites/buildings listed for their architectural merit, etc)?

2 Hazardous substances

- Do you store/handle/use any hazardous or potentially hazardous substances on site?
- If you are storing hazardous substances on site, do your local emergency services and planning authority need to know?
- Have you identified hazardous substances through your current management of health and safety issues?
- Do you have the appropriate information concerning the use, handling and eventual disposal of these identified substances?
- If using below ground storage tanks, how do you inspect them and any pipework for integrity?
- Are any of the substances potential pollutants mentioned in regulations covering disposal?
- Do you manufacture or use ozone depleting substances?
- Do you have any sources of radioactivity on site?

3 Air

- Do you manage any processes, operations or substances that give rise to emissions to air (include vehicle exhaust emissions from transport)?
- Do you know what they are in detail (type/amount/circumstances of release)?
- Do you have an up-to-date site plan showing air emission points?
- Do you monitor any of those emissions? Why?
- Do you specifically monitor greenhouse gas emissions? (For use in an emissions trading scheme?)
- Do you have any technology installed to prevent and/or minimize harmful air emissions?
- Do you know who your regulator of air emissions is? Do you know the officer by name and have you spoken to them recently?
- Has the regulator had any complaints about air emissions and/or odours? If so, how did you respond to their communications?

4 Water

- Do you manage any processes, operations or substances that give rise to discharges to water (include storm water management, discharges to sewers and controlled waters)?
- Do you know what they are in detail (type/amount/circumstances of discharge)?
- Do you monitor any of those discharges? Why?
- Do you have any technology installed to prevent and/or minimize harmful discharges to water (include water treatment plant)?
- Do you have an up-to-date site plan that shows the current discharge points and drainage routes?
- Do you know who your water discharge regulator is? Do you know the officer by name and have you spoken to them recently?
- Has the regulator had any complaints about your water discharges? If so, how did you respond to their enquiries?
- Do you have any requirements to conserve water designated by the appropriate regulator?

5 Noise

- Are you in a residential area?
- Do you manage any processes, operations or substances that give rise to noise and/or vibration (include, particularly, normal noise levels during late shift work and vehicles on site)?
- Do you monitor noise in the workplace environment for health and safety reasons? If yes, have you got access to that information in detail?
- Do you monitor any other external environmental noise? Why?
- Does your land use/planning permit stipulate any conditions concerning noise?

- Do you know who your regulator of noise is, and what the regulations are? Do you know the officer by name, and have you spoken to them recently?
- Has the regulator had any complaints about noise from your operation? If so, how did you respond to their enquiries?

6 Waste

- Do you know how the appropriate regulator classifies different waste types?
- Do you know how many of those different waste types/streams you have within your operation?
- Do you segregate and/or monitor your different waste streams?
- Do you keep detailed records of waste disposed of from your site?
- Do you have the appropriate paperwork (permits/consents/licences) associated with each of the waste streams?
- Do you dispose of waste on site?
- Do you treat waste on site?
- Do you have a site plan that identifies waste storage points?
- Do you have any reuse or recycling schemes in operation (include those that apply to your products as well as your processes)?
- Do you know how much and what type of packaging material passes through your site?

Do you know how to structure your monitoring of legal developments?

Once you have answered the questions listed above, and have compiled the detailed information required, the scope of your legal monitoring requirements will become much clearer. However, the structure of that monitoring and the associated procedures that you employ will need to be further refined to ensure effective and efficient coverage.

Though electronic or paper-based databases of environmental law are popular, it is not enough to simply subscribe to one of these commercially available tools. The information contained in any database is useless unless monitored, evaluated and communicated to the relevant personnel. Responsibility for this procedure should thus be clearly defined, allocated and supported by a carefully thought through procedure. It can be split between managerial and operational personnel, though if this is your preferred option, care should be taken to avoid confusion. The main purpose is to ensure that the organization's environmental performance stays within legal requirements.

As a basic overview, a procedure for these activities should attempt to address the following:

- Who is going to monitor your legal requirements?
- How will they know what to monitor?
- How will they assess the information?
- Who will sign off the completed updated material?

PLANNING

- Who will need to know details of the new information?
- How will they get this information?

Legislation at international, European and national level moves at relatively different speeds, as does any change in guidance to the way the legislation is realized at ground level through regulations. The closer the issuer of the legislation to your organization, the faster the changes will come through. International agreements can take many years to make their impact on the law of any country, and European Directives can take up to five years before their effects filter down to be felt at the level of a particular site. (Remember too, that even though EC Directives need national implementation, however long that may take, European legislation does have primacy in European Member States.) On the other hand, it may only be a matter of months for a local regulator to change their guidelines and the way they apply any of the regulations in their remit. It pays to stay in close and regular consultation with your regulator(s), as they will ultimately have the most realistic idea of the timeframe of any developing laws.

It may help your monitoring of the law if you divide the legislation that you are keeping an eye on into 'Levels of interest': Level 1 being the long-term developments over, say, the next five years, down to Level 6 which may be monitoring minor changes as close to you as five weeks. Your grading system might end up looking something like this:

Level 1: International agreements (Montreal Protocol, Kyoto Climate Change Protocol, etc)
Level 2: European level (Directives and Regulations)
Level 3: Primary national legislation
Level 4: Secondary national legislation (Regulations), Regional/Federal Government legislation
Level 5: Guidance from regulators/inspectors
Level 6: Rumours, gossip, speculative journalism, passive smoke breaks

Do you know which elements of your EMS the law affects?

Policy issues
(ISO 14001 Clause 4.2 /EMAS I-A.2)
The EMS standards include the identification of relevant legislation as a key part of the system, with EMAS going as far as a requirement for a register of the appropriate laws to be compiled. In either case, a procedure to continue the monitoring of relevant legal developments will ensure that the activities, products and services affected are managed in such a way as to fulfil the policy commitment to abide by the law. Both EMAS and ISO 14001 (and thus all the phased EMS implementation schemes), in defining the management system in such a way as to link the entire system with the organization's policy, ensure that the commitment to obey the law contained in the policy is therefore reproduced throughout the other elements of the system. Below are more detailed examples of how that works.

Legal and other requirements
(ISO 14001 Clause 4.3.2 /EMAS I-A.3.2, Article 6(4))

Once you have a system for identifying the law and keeping it up to date, a manager must still ensure that the procedure is effective (ie that the information gets to the right people and is accurate). It is certainly true that a failure at this level may lead ultimately to a failure to comply with the law, but procedural failure does not automatically imply legal failure. For example, if the EMS has identified that a water discharge consent is necessary from a regulator, then that consent may also contain specific conditions (such as, say, temperature or pH level) that will require monitoring and recording. Obviously, if the organization does not recognize that it should have a consent in the first place, then it may follow that it will not be monitoring the appropriate criteria. An auditor, internal or external, may choose to record such a non-conformity under ISO 14001 Clause 4.5.1 (monitoring and measurement), but this would not be accurate. It would fail to have discovered the root cause of the problem (not identifying the correct legislation), and have simply given a superficial record of the symptom (insufficient monitoring).

There is a tendency among organizations setting up an EMS for the first time to think that all non-conformities of the system that include a legal aspect can be traced back to these clauses of the standards; the identification and monitoring of relevant environmental legislation. That misses the point. Such a diagnosis ignores the more fundamental legal influences on other elements of the EMS listed here. Because there is a history of many organizations already indulging in legal compliance audits (at whatever level of adequacy), there is a strong tendency to react to such a situation by treating it as a de facto non-conformity with the requirement to identify the appropriate law. This does not allow them to look deeper and find a specific breakdown in other parts of the EMS itself.

The clause also refers to 'other requirements' which is easy to overlook when it comes to scoping this area and any associated procedures. 'Other requirements' refers to voluntary requirements to which the organization or site subscribes, such as codes of practice, industry schemes, or even environmental dimensions of customer specifications and so on. It is worth noting that this last factor can be an additional key part of the inspection criteria of phased EMS implementation schemes; they can include external monitoring of contractual environmental requirements.

Objectives, targets and programme(s)
(ISO 14001 Clause 4.3.3 /EMAS I-A.3.3, II(2), I-A.3.4)

The link to legal compliance in this area may not be immediately obvious, but the effect is still there. Objectives and targets should reflect the organization's policy, and as the policy should also contain a commitment to obey the law, the objectives and targets themselves should be set within legal parameters. Not only should any specific impact or aspect mentioned in the policy have a matching objective and series of targets, the objectives themselves have to recognize compliance with the law as a minimum. One of the reasons that the system has been written in this way is to make it impossible for an organization to set itself an objective that is less than the legal requirement and remain in conformity with ISO 14001 or EMAS.

In the same manner, the law pervades the management programmes which are designed to deliver those objectives and targets. Again, it is impossible for such a programme to be designed to deliver performance or management objectives below the limits required by the law, and yet remain in conformity with the EMS standards.

Competence, training and awareness
(ISO 14001 Clause 4.4.2 /EMAS I-A.4.2)

If an organization is attempting to maintain legal compliance, then it follows that it is important that the relevant personnel know and adhere to the law covering environmental aspects of their work. Again, using the relevant legislation as a minimum requirement, a competency assessment and any training and awareness programme should reflect this relationship. On occasions, there is a tendency for new EMS systems to concentrate this awareness and training only on those personnel who have responsibility for areas where there are significant impacts on the environment. Though this is the minimum that should be aimed for, to restrict the training to these staff members would still be a non-conformity and could be less than is required to ensure legal compliance. For example, training in emergency response should be broadened to include all personnel, as should a specific programme that allows all personnel to identify the problems of departing from known procedures.

Operational control
(ISO 14001 Clause 4.4.6 /EMAS I-A.4.6)

If the legal requirements placed upon an organization are not incorporated into the design of the EMS and its procedures, there is every likelihood that the operational control required will not be evident, and a breach of the regulations will occur. For example, if an organization was required by law to monitor its emissions of a particular substance, and the appropriate monitoring equipment was installed, calibrated and maintained, yet there was no written procedure to cover the operation of the process that produced the emission in the first place, then there is not sufficient operational control being exerted. The non-conformity to the standard would lie in the lack of a procedure, but it would be all the more serious because the process produced a significant environmental impact already identified, not only by the company, but by an external regulator.

Evaluation of compliance
(ISO 14001 Clause 4.5.2 /EMAS I-A.5.2)

With or without an EMS, companies must comply with legislation. Obviously, complying with the law is not a one-off exercise but a continuous process. With increasing levels of legislation, it will come as no surprise that the periodic evaluation of the level of organizational compliance has become a key factor in installing and maintaining an effective EMS. Early experience of EMS development showed that, although compliance audits could be built into the scope of EMS internal audits, a separate and distinct 'legal compliance only' exercise was far more likely to ensure that a complete and detailed picture of compliance was transmitted to the management. If there is an oversight in the original identification process that we are talking about in this chapter, it is true that the periodic evaluation process might pick up the gap at a later stage. If this happened, however, there is still no guarantee that a legal problem has not emerged in the interim and it is therefore not a wise move to rely on a later audit to make up for any initial shortcomings in the legal identification process.

Non-conformity, corrective action and preventive action
(ISO 14001 Clause 4.5.3 /EMAS I-A.5.2)

Again, wherever the law is identified as relevant to an environmental impact, it has the effect of magnifying the seriousness of any non-conformity to the standard. In the case of this particular EMS element, lack of a corrective action on an identified non-conformity (discovered during, say, an internal audit) would not only indicate that the organization is not following its own procedures, it would also expose the organization to a much increased risk of breaking the law. EMAS achieves much the same effect by ensuring that the definition of the phrase 'environmental audit' focuses the audit process on evaluating compliance with 'company policies'.

Control of records
(ISO 14001 Clause 4.5.4 /EMAS I-A.5.3)

Many regulations and legislation in the environmental area require that specific records be created and maintained. In such cases, it is particularly important to check to see if the regulation stipulates a retention time for any records produced. It follows that any procedures that the organization uses in conjunction with these records should reflect the legal requirement as a minimum.

Extra EMAS element

The revision of EMAS in 2001 did away with all but the most necessary differences between ISO 14001 and the regulation. However, those using EMAS need to be aware of the ongoing close relationship between the regulation and the law. Once a site has produced an environmental statement, had it verified and had its environmental management system validated, the Competent Body responsible for running the scheme at a national level can accept the site onto its EMAS register. Organizations can also be removed from the register should a regulator inform the Competent Body of a breach of any relevant regulations (EMAS Article 6.4), thus reinforcing the close relationship between the law and the continuing operation of the management system.

Do you know where to go for further information?

Sources of information concerning environmental law are not limited to those employed directly by the legal profession. Approaching the regulators themselves can often be helpful in seeking clarification about what applies to your processes, products and services. Not all regulators are as Draconian as they may seem, and many respond positively to enquiries seeking further information, as they regard it as part of their remit to produce and perhaps even clarify their own guidance notes on the application of the law and related issues. However, responses by regulators can vary according to individual officers and the legal framework within which they work, so approach with preparation and caution.

Some organizations may feel that their relationship with the regulator is potentially too sensitive to allow such an open approach. In these cases, there are still other avenues that can be explored. With the advent of EMS standards, many legal practices, environmental consultancies and information providers have collaborated in the production of legal databases focused on environmental law. Such databases can be produced in paper form, or in CD-ROM formats which allow the user to search quickly for relevant information. In a similar vein, it is often worth exploring what information may be available to you and your organization from the Internet (including the World Wide Web).

Professional bodies and trade associations exist in order to maintain standards of performance and ethics among their particular members in a specified industry. You may find that at a national level there are several in the areas of environmental auditing, environmental management and environmental consultancies. Many will produce a prospectus for potential members and it is worth reading a selection of these in order to choose the most appropriate for your own circumstances. Such bodies may also maintain a database of members which makes it possible for them to operate a referral service, either to acquire those services directly, or simply to put you in touch with companies in a similar situation who may be willing to share the information they have about the law. Most bodies also have journals which frequently survey the development of environmental law, with the best publications giving both an overview of legislation as a whole, and in-depth analysis of the interpretation and application of the law. These and other commercially available publications can also often be the best source of material concerning the long-range development of international and European level legislation.

Local business support organizations, such as regional government offices, chambers of commerce, business clubs and similar, are now beginning to realize that environmental concerns are not limited to specific companies, but apply across the board. The most responsive of these organizations are already aware of the problem of both tracking and understanding the complex issues covered by environmental law, with many arranging forums and seminars on the subject. If your local organizations have not yet thought about this area, it may make sense to suggest it and initiate something with like-minded companies.

Do you know what is meant by 'other requirements' in EMS standards?

Both ISO 14001 (Clause 4.3.2) and therefore EMAS (I-A.3.2) do not restrict organizations to assessment of legislation when planning their management systems. Both make references to 'other requirements' which may have a role that is similar to legislation. Customer (second party) audits can also include contractual environmental requirements in their scope and may even be included as part of a client-supported phased implementation scheme. All these requirements may also affect all the elements of the system in a similar way, and will certainly be an important driving factor in providing baselines beneath which operations cannot occur. Such 'requirements' may include:

- internal performance standards;
- client requirements;

- relevant clauses of quality policies;
- relevant clauses of health and safety policies;
- site environmental policies;
- corporate environmental policies;
- trade association / industry sector initiatives (sector guidelines, etc);
- national/regional/local initiatives (waste minimization schemes, etc);
- codes of practice;
- international agreements and guidelines to which the organization subscribes (Global Reporting Initiative and similar).

 Narratives

Smallco

Alongside establishing significant environmental impacts, the IER scope included an appraisal of relevant environmental legislation. Clearly, much was identified as relevant and a first draft legislative requirements grid has been drawn up. However, the exercise suggested major gaps in Smallco's understanding of environmental legislation. They have budgeted to buy in some legal expertise. The reality is that the advice from external colleagues and estimates from two local law firms indicate that Smallco need the advice but cannot afford the costs. The MD, who acknowledges the importance of the work simply cannot commit extra budget. He has asked the Ops Manager to prepare some ideas as to how the legal knowledge gaps might be filled at low cost and the help offered by the customer company might well plug the gaps. As part of their phased EMS implementation scheme, their 'mentor' client company helped to a degree, allowing them sight of their basic legal identification procedures and signposting them to some potential support organizations, but site and process specific information proved hard to identify.

The Ops Manager prepared a handout (Table 5.1) to aid discussion.

Another phase is now completed in the EMS implementation. The draft phased implementation plan is providing an important point of reference to measure progress against.

BIG Inc

For an IT company, using a commercially available environmental law database seemed the ideal solution. The existing IT network meant information could be disseminated easily enough, and updated at regular intervals. What soon became clear, however, was the lack of in-house expertise to read, analyse and categorize the information. Having other European offices, spread across five different countries, also meant that a huge amount of information would have to be examined on a regular basis.

In the end, after consulting with the director who had been appointed as the EMR, the project manager gave the job to the Services and Facilities Manager. He already carried the responsibility for reviewing health and safety regulations for the company headquarters. Expanding his remit in this way had a knock-on effect on his health and safety work which also had to be expanded to cover the whole organization (previously business managers in

Table 5.1 Filling legal knowledge gaps

Option	For	Against	Action?
Seek support from major customers involved in supply chain initiative	Low cost improved relationships	None	
Invite the regulators in to discuss and advise on requirements	Low cost and potential endorsement of work	Might reveal non-compliance areas and expose Smallco	
Consult trade associations/local business support services	Low cost, good industry focus and networking	This broad brushstrokes approach might miss some legal issues	
Subscribe to a generic legal updating service	Medium cost and hopefully all info. will be provided	Time and responsibility for someone to derive a relevant listing which may still have gaps	
Subscribe to law firm's environmental legal briefing services	Medium cost	Again might be too generic and not close all gaps	

each country had had the responsibility). His first scoping of the job revealed that focusing on European directives at an early stage helped to identify which later national legal developments should be monitored, a task that prompted the software design team to start looking at adapting their existing knowledge management tools to help with the job.

As a 'safety net', the project manager talked to the company's external legal advisers who agreed to check through the material once it had been compiled, and to do so for the first year at least, until the system was more established.

Trade secrets

- If you are a smaller site, or part of a larger organization, then find out what corporate legal services may be available to you.
- Whether or not you are taking part in a phased implementation scheme, you could always approach friendly suppliers/customers for information on how they approached the legal identification process.
- Contact local companies of a similar size, or with similar processes. They will have the same local regulators, and may have already done some work that they are prepared to share with you. Local business clubs are an ideal place to make such contacts.

- When deciding which legislation to monitor, try to grade it in terms of its immediate relevance to your organization's core activities, and split the monitoring into short, medium and long term. Monitor the most relevant/closest to your operations most frequently.
- When designing a procedure to identify and monitor legal and other requirements, don't forget to include how material will be updated, who will use the information and how will they get access to it.
- When compiling a record of relevant legislation, remember that most company directors are legally responsible for the activities of their company. It may be appropriate to have a top management representative (a board member) sign off the final report or documentation to indicate that they understand and agree with the contents. This will reinforce environmental issues as being a board-level priority.
- Remember that some new legislation may apply to your organization on a retrospective basis, or may be affected by past activities, products and services. It's easy to forget that the past can be affected by the present in terms of new legislation.

Things to think about

- It is always difficult to predict accurately exactly how environmental legislation is likely to develop in the future. However, both internationally and at European level, certain key areas of discussion will undoubtedly influence national legal frameworks:
  - managing the environmental impacts of products and associated activities and services (such as manufacturing and transport), by taking a life cycle approach;
  - the efficient use of resources, which may be termed as 'moving towards sustainable consumption and production';
  - creating and delivering biodiversity action plans;
  - providing evidence of practical corporate and social responsibility.

PLANNING

6

Developing a Policy: Where do We Want to Go?

An environmental policy is a short written statement of intent which sets out your organization's position on the environment and is consistent with the scope of its EMS. It is the driver for both initial implementation and ongoing improvement of the EMS. It reflects top management commitment and is the foundation for internal and external communication and the setting of objectives and targets which define the implementation and running of the EMS.

✓ ISO/EMAS quick check

Area of EMS	ISO 14001	EMAS
Environmental policy	4.2	Annex I-A.2
Legal and other requirements	4.3.2	Annex I-A.2
Objectives, targets and programme(s)	4.3.3	Annex I-A.3.3
Competence, training and awareness	4.4.2	Annex I-A.4.2
Communication	4.4.3	Annex I-A.4.3
Documentation	4.4.4	Annex I-A.4.4
Operational control	4.4.6	Annex I-A.4.6
Evaluation of compliance	4.5.2	Annex I-A.5.2
Management review	4.6	Annex I-A.6

Chapter executive summary

Move to the next chapter when you can answer all of the following questions in ways which make sense to you and your organization:

> *What is an environmental policy?*
> *When do I need to write one and how should I write it?*
> *How do I design an effective policy?*
> *What will I need to include in it?*
> *What needs to be done with it in the context of the EMS?*
> *How will I communicate with everyone?*

You are here

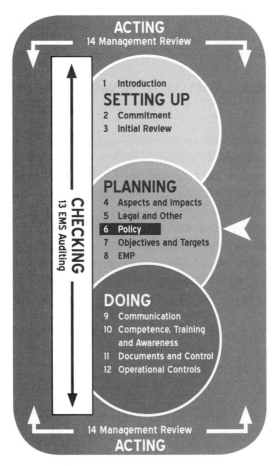

Toolkit requirements

To work through this chapter you will need the following knowledge and skills:

Knowledge

- the outcomes of and understanding from the IER;
- the management culture;
- the key stakeholders and interested parties;
- the business setting: internal and external including long- and short-term business issues;
- available resources;
- the level of general environmental awareness;
- dissemination routes.

Skills

- writing/drafting skills;
- communication skills, both verbal and written.

Context

As well as being a requirement of the standards there are clear benefits in setting out your business strategy for environmental management in a policy statement:

- Your policy is a commitment to improve performance continually in ways which complement the organization and is in tune with other business objectives.
- Your policy tells stakeholders – identified in the IER (Chapter 3) – that you are managing your business effectively and begins the process of managing the relationships to mutual advantage.
- The policy establishes the template for objectives and targets (Chapter 7).
- The policy sets internal standards for the EMS and expectations of your staff.
- The environmental policy is a formal resolution to manage those aspects of your business which have an environmental component. A business- and environment-focused policy will bring cost savings, increase compet- itive edge and numerous bottom line business benefits. The policy demon- strates proactive commitment.

Tasks

What is an environmental policy?

In many ways, the environmental policy is much more than a short written statement of intent which sets out the organization's position on the environment. It has three key func- tions that affect the EMS at a fundamental level:

1. It identifies the main issues for the business.
2. It sets out the broad brushstroke actions that will be taken to manage these issues to make business sense of the environment.
3. It signposts actions that will be taken to both monitor and improve the organization's environmental position.

The EMS standards and related implementation schemes clearly establish the areas an environmental policy should address. These requirements are set out in the 'Quick check' section at the beginning of this chapter.

Clearly, to establish and manage an effective EMS the policy is much more than a written exercise in simply meeting the requirements of the standard(s). For the organization the environmental policy needs to reflect the realities of the wider environment in which it operates. The well-established business planning tool of PEST (**P**olitical, **E**conomic, **S**ocial and **T**echnological) analysis can help provide this overview. For environmental manage- ment it is worth adding a final 'E' to make PESTE. The 'E' signifies Ecological and establishes the important linkage to the external environment. To unpack 'PESTE' a little more you can ask yourself:

Political	What is the political climate in which the organization operates? How do we keep track of political changes?
Economic	What are the economics of our business? What do we do well and how does this meet the bottom line?
Social	What social considerations are there? What stakeholder pressures exist and how do we deal with them?
Technological	How do we operate our business? What is the appropriate use of technology and how do we make technological investments?
Ecological	What do we know about the environmental aspects and impacts of our business (see Chapter 4)?

When do I need to write one and how should I write it?

The elementary answer to this question is: 'When it makes sense to'. A policy written to satisfy an entirely externally imposed agenda is unlikely to give the organization the kind of flexible, proactive, clear and simple match to its own needs. As a medium-term consequence, it is unlikely to be sustainable either. It is important to generate an environmental policy as part of the wider EMS development strategy. If the policy is to provide the right foundation for an effective EMS, then it needs in turn to be founded on:

- business-focused data within the scope of your EMS;
- a good understanding of EMSs, Chapters 1 and 2;
- an environmental stock take of the present situation, often called the base-line or initial environmental review, Chapter 3;
- a thorough understanding and knowledge of the organization's environ-mental aspects and impacts, Chapter 4;
- an understanding of legal and other requirements pertinent to the organ-ization, Chapter 5.

The policy must mirror organizational reality and anticipate the future too.

How do I design an effective policy?

For the environmental policy to work in practice there are five key areas to address which you should incorporate in your design. These are outlined below.

Attitude

To establish an effective policy everyone in the organization needs a positive environ-mental attitude. Commitment and responsibility make the glue that holds the policy together. Involving relevant staff in the drafting of your policy – from line managers to shop floor representatives – and ensuring that they understand why the policy is being produced and its implications for the business are vital tasks. To satisfy the standards requirements (if you are pursuing them) and ensure that the policy and the EMS it represents are rigorous, and then ensuring that the final version is approved and agreed by senior management, is essential.

Accuracy

The environmental policy needs to be a marker on the journey not a millstone around the organization's neck! It needs to set out goals in an open, accurate and realistic way. The recipients of the policy, whether customers, contractors, suppliers or staff, will need to be satisfied that what has been set out is realistic for the business and for them.

Adequately resourced

Resources will be required to move the policy from paper intent to active performance and amplify the efforts of those involved in the implementation of the EMS. Resources will include both staff time and money. The mix and allocation will depend on identified business priorities.

Awareness

The policy and the actions it suggests will need to be communicated. Effective communication, both within the business, and outside the business is essential. The actions of staff to bring the policy to life will bring credibility and credit to the organization. Staff will need help and guidance to integrate the policy with their own workplace activities. To maximize the benefits of your new environmental attitude, anyone working for or on behalf of the organization will need to know what is now expected from them. On a wider front other stakeholders such as the community – from the local to the national or perhaps international – will need to know about the new position.

Action

Environmental management is a journey and not a destination; the policy should reflect this by indicating how the organization will monitor progress, audit the EMS and act proactively.

What will I need to include in it?

The policy should include a set of specific statements for your organization. There are five key areas to consider in the draft:

1   A very brief overview about what the organization does in terms of its activities, products and services.

2   A broad statement of intent. This should outline both the role of the business and how it proposes to maximize the opportunities presented by managing the environmental aspects of its activities. For example:

> The Organization Ltd is committed to design, develop, produce and deliver quality goods and services. We recognize that day-to-day operations impact on the environment in ways which are both positive and negative. We wish to minimize the potential harmful effects of such actions wherever and whenever this is practicable and work to secure measurable business benefits from our EMS.

3   Statements on specific issues. These are likely to emerge from your work in Chapters 3 and 4 and will include 'in principle' statements on various issues such as:
- how the business will address stakeholder issues;
- how the business will ensure legal issues are addressed;
- how the business will deal with key environmental issues identified as significant and associated with its operations with an emphasis on preventing pollution;
- how so-called 'soft' issues such as training and communication will be dealt with.

4   A statement of what is expected of external parties. Many external organizations will be associated with your organization including customers, suppliers, vendors and contractors and the policy should commit you to communicate with them.

5   A statement indicating how both management systems and performance will be developed and continually improved. Without a commitment and action to implement, review and update the EMS in order to achieve improvements in overall environmental performance then fine words could well be the end of your involvement with environmental management. The policy should commit the organization to these activities.

The final balance of the policy will depend on those environmental issues you identified as priorities for your business. An example policy is shown in Box 6.1.

Box 6.1 Best Foot Forward environmental policy

Best Foot Forward is committed to reducing its impact on the environment. We take our responsibilities seriously and pursue a policy of environmental best practice with clear policy aims and objectives.

Policy aims
- To include environmental considerations in daily and project activities.
- To conduct an ongoing company resource flow analysis and ecological footprint.
- To implement an environmentally responsible purchasing policy.
- To reduce, recycle and reuse waste on the premises and at BFF events.
- To maximise energy efficiency and reduce greenhouse gas emissions.
- To minimise and control the use of water on the premises.

Objectives

Energy
BFF aims to maximise energy efficiency and reduce emissions by:

- minimising and monitoring its total energy consumption;
- purchasing, where possible, renewable energy.

Purchasing and supply

BFF aims to maximise the use of environmentally sound products and services, by:

- maintaining and increasing the use of local suppliers for products, such as stationery;
- using environmentally responsible and recognised printing companies for company publications.

Waste

BFF aims to maintain and improve its management of waste produced on the premises by:

- minimising (reducing and/or recycling), wherever possible, its solid and liquid waste streams;
- using recycled products wherever feasible;
- adopting a purchasing policy sensitive to environmental concerns;
- ensuring satisfactory disposal of waste that cannot be re-used or recycled.

Transport

BFF aims to maximise the use of efficient and environmentally sound transport options by:

- minimising the use of cars and aeroplanes for transport;
- maintaining and improving use of public transport for business activities;
- maximising the use of environmentally sound transport options for both project-related work and commuting.

Water

BFF aims to manage its water resources efficiently by:

- minimising and monitoring the total water consumption;
- ensuring that water systems on the premises are not wasteful.

Based on the findings of our environmental reports, Best Foot Forward sets targets and priorities for action for the following year.

Health and Safety

A health and safety policy has been drawn up for Best Foot Forward, to which the environmental policy is a component. The office is designed to ensure a pleasant work environment in which staff can interact with one another, yet remain work-focused. Best foot Forward also has a no-smoking policy.

Source: The above environmental policy is reproduced with thanks to Best Foot Forward who were awarded the Queen's Award for Enterprise: Sustainable Development 2005. www.bestfootforward.com

In formulating an environmental policy it is important to strike a balance between long- and short-term issues. The payback for some policy activities may be relatively easy to measure, others may be less so; for example, an investment in new machines might take several years to pay back while energy conservation measures or waste minimization programmes can bring an almost immediate return. Sustaining a viable market share or satisfying local or national stakeholders' expectations or demands are much less easy to measure in hard fiscal terms although techniques and tools for doing so are rapidly developing. The strategy which is embodied in the environmental policy needs to reflect these changing factors in the business environment and be flexible enough to be able to respond positively in ways which secure business benefits. It is well worth getting hold of a wide variety of policies from other companies – not just those from your own sector – and see if there is anything you like about them and if there are any elements which might help in your own policy drafting work.

Drafting the policy

The environmental policy is a cornerstone of the EMS. At this stage you should be ready to begin to assemble a first draft of the policy. It is well worth preparing a first draft for internal consumption only. Make sure you mark all copies up as 'Draft only' and 'Confidential' – if you don't they have a habit of leaking outside before you are ready. You will have assembled a team of trusted colleagues working with you to develop the EMS. They will have been closely involved in the IER and setting of objectives and targets which underpins the policy commitments, so make use of their expertise for critical comments on this first draft. It is worthwhile adding a couple of neutral readers to the list of critics, for example, a trusted colleague from another organization who will be able to take a disinterested, external view of the draft. Their observations can be a vital component in developing a policy which will make sense to a wider audience.

With this expert input, redraft the policy to produce a presentation draft for discussion with senior management. It makes good sense to discuss this draft policy alongside your emerging objectives and targets for the EMS. Remember, the policy provides the framework for setting objectives and targets so the final sign off will come when all are satisfied that the policy and the objectives and targets are working in tandem. Finally, produce the first version of the organization's environmental policy and make sure it is a controlled document and signed off by senior management.

What needs to be done with it in the context of the EMS?

There are several things to do with the new environmental policy. Some, such as making it available to the public, are requirements of the standards. Other actions are fundamental to making the policy an instrument of the EMS. Use the checklist below and consider each point in turn.

- Make sure you send a copy to everyone who helped in the consultation process leading to the drafting of the environmental policy.
- Distribute a copy of the policy statement with a suitable endorsement from senior management to all employees. This could be by paper or electronic means.
- Attach copies of the policy to workplace notice-boards and in staff handbooks.
- Publish the policy in company reports and other relevant communications.

- Ensure staff awareness of the policy by including it in all relevant training activities.
- Be proactive. For example, let your customers and suppliers have a copy of your policy and let them know your growing environmental expectations of them.
- Measure the impact of the policy in the workplace to ensure that the message is being received, understood and acted on.
- Keep it up to date. Elsewhere, your EMS will have committed you to continual improvement and this will mean a review of the policy at determined intervals.
- Let staff know how the organization is performing against its policy objectives.
- Match your distribution of the policy to the number and type of stakeholders you identified in Chapter 2. How have you covered each stakeholder group? Think especially about suppliers and contractors whose activities may relate to your significant environmental aspects (Chapter 4)

 Narratives

Smallco

Smallco are now confident that their EMS is on track and tackling it in a phased way has helped hard-pressed managers by not overwhelming them in trying to do everything at once. The MD is pleased that EMS work has complemented and not displaced core business activities. They have now written and internally published their environmental policy. The policy has been through two internal drafts and refined into the finished product. Help and advice as part of an ongoing supply chain initiative where larger customers are helping smaller suppliers in the implementation of environmental good practice has been invaluable and helped forge better working relationships. The policy commits Smallco to communicate its contents to the stakeholders established at the baseline assessment stage. The options for disseminating the policy, and therefore Smallco's expectations, are discussed at one of the monthly management meetings. The MD sums up the discussions into four action steps:

1 All relevant stakeholders – as identified in the baseline assessment – will be sent a signed copy of the environmental policy.
2 The Ops Manager will provide a couple of generic paragraphs for inclusion in the covering letters.
3 Each manager, including the MD, to prepare individual covering letters, dated the first day of next month, for stakeholders they deal with on a regular basis.
4 The MD's secretary will compile a master mailing list which will be checked by the Ops Manager to ensure that all stakeholders are included. Reply forms, with space for their suggestions, are included for customers, suppliers and contractors.

BIG Inc

Writing the policy for the EMS turned into a longer process than anyone had originally expected. Defining the scope of the EMS was a major challenge for an organization with nine UK sites alone. Though the board were quite certain of their views on the subject, they wanted it to be dynamic and in line with other parts of the company's business culture. The real problem here was the nature of the environmental aspects and associated legislation that the system would have to tackle. Because most of the impacts were indirect, requiring the exercise of influence rather than control, the policy had to recognize that the time-scales for many of the initiatives would be long with 'soft' objectives. As a result, early drafts of the policy were turned down by the board as 'not dynamic enough'. Even when the EMR attempted to write a statement, they felt it failed to capture the imagination of the reader.

Eventually, at the suggestion of the Chief Executive, the board undertook a facilitated brainstorm exercise, and decided that they could sell the environmental aspects of their work to their clients as part of their added-value consultancy. This decision opened up the potential development of a whole new strand of business, specifically advising clients on how they could improve their environmental performance by using BIG Inc's IT solutions. The idea certainly helped to bring on board the Marketing Director, who then found it relatively easy to draft an EMS policy aligned with the rest of the company's values.

Trade secrets

- Consult widely for feedback on the draft, and make sure those who will be affected by it are involved and not surprised.
- Make sure the policy is based on your organizational situation and the scope of the EMS you have carefully defined, not someone else's policy which looked good enough to copy!
- Don't make promises you cannot keep or suggest objectives and targets you cannot work with. A 'save the planet' policy will undoubtedly backfire on the business.
- Keep it simple. The policy is not your EMS, just part of it. One side of paper provides plenty of space for an excellent policy.

Things to think about

- As more and more EMSs become mature, larger organizations will find themselves working more closely with their smaller suppliers. These supply chain forces will result in more effective integration of environmental policies and a shift towards an overall compatibility of message.
- Businesses are finding it makes good business sense to bring together quality and health and safety with environment, developing one overarching policy to address all three areas. Expect to see, however, that this is revisited in the light of sustainability and governance concerns. Environmental policies will become a key factor in guiding strategy and business development, elevating it from the current role of operational post hoc application.

7

Developing Objectives and Targets: What Are the Landmarks on Our Route?

Setting viable objectives and targets is not always as easy as it might appear at first glance. Not only do they have to be in line with your previously drafted policy (including the commitment to continual improvement if you are pursuing formal standards), and your already identified significant impacts, they should be specific, measurable, achievable, realistic and time-related. Effective objectives and targets are the only way to set up effective delivery programmes (see Chapter 8). The policy should have outlined the methodology of the organization's approach to the environment. Objectives and targets will then give such intentions greater relevance on a day-to-day basis for the organization and all levels of the workforce. If you are involved in a phased implementation scheme, you may already have undertaken a simpler form of this exercise in the opening phases. You will still need to revisit your original 'quick and dirty' objectives at some point as your EMS matures, so the information in this chapter is relevant.

✓ ISO/EMAS quick check

Area of EMS	ISO 14001	EMS
Environmental policy	4.2	Annex I-A.2
Objectives, targets and programme(s)	4.3.3	Annex I-A.3.3
		Annex I-A.3.4
Monitoring and measurement	4.5.1	Annex I-A.5.1
		Annex II-2

🗍 Chapter executive summary

Move to the next chapter when you can answer all of the following questions in ways which make sense to you and your organization:

> *How are objectives and targets set?*
> *What is the relationship between an objective and a target?*
> *Are there different types of objectives in an EMS?*
> *How is an objective made effective (how do I know I have set the right one)?*
> *Can EPIs help to set targets?*
> *What other elements of the EMS are affected?*

⌇ You are here

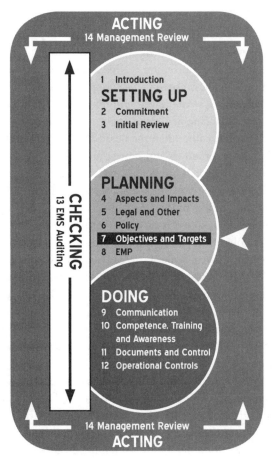

⌇ Toolkit requirements

To work through this chapter you will need the following knowledge and skills:

Knowledge

- company policy commitments;
- business plan/objectives;
- potential resource allocation;
- views of interested parties;
- accrued benefits from planned improvements;
- relevant current and future legislation;
- EPIs.

Skills

- information analysis;
- liaison skills;
- gaining support/obtaining buy-in from others.

Context

The setting of objectives and targets can make or break an EMS; it can turn rhetoric and theory into basic shop floor practice, or it can impose arbitrary burdens on hard-pressed personnel, who may do everything to downgrade the imposition of such irrelevant targets. The difference between success and failure in this is not always immediately apparent, simply because objectives can appear reasonable enough on paper, only proving difficult or impossible to achieve after the system has been running for a year or two.

It is a continual balancing act, since environmental objectives, like politics, can become the art of the possible, rather than the desirable. Business and financial objectives may take first priority under certain circumstances, and these will vary from organization to organization. Financial projections are not always accurate, resources become squeezed, markets expand and contract, the flow of orders may speed up or slow down, regulatory guidelines may change and in the middle of all these variables, the EMS manager has the unenviable task of setting even more objectives and targets for the hard-pressed workforce.

Even when the objectives have been set, individual 'stepping stone' targets also have to be worked out in order to make progress towards the ultimate goal measurable over time. At the end of that process, resources have to be allocated, and the objectives have to be accepted by those in the positions of responsibility. It is vital to get what has been called 'an alignment of personal and corporate objectives' if environmental performance overall is to improve and benefits accrued. Many employees are suspicious of a new set of objectives, viewing them as another unwelcome series of personal rather than environmental performance indicators.

If the organization has already had recent experience of other management systems being introduced, such as quality or health and safety, it would be hardly surprising if staff were to react with 'initiative fatigue'. The way to achieve the desired alignment is through specifically focused training that takes into account the levels of awareness needed by individual members of the workforce (see Chapter 10). Even the most carefully thought through objectives can come unstuck if personnel either do not understand, or do not want to support the goals.

An alternative approach used in phased implementation schemes is to concentrate in the early stages of development of the EMS on environmental performance evaluation techniques. From the very start, organizations are encouraged to manage their environmental impacts by using a series of performance indicators as a feedback mechanism. This can work well for the first few cycles and has the advantage of being easy to follow as well as set up. However, over time, as the early gains are achieved (ie cutting energy bills or reducing specific waste streams) even these performance indicators will need to be revisited requiring more complex responses and thus objectives and targets from the organization.

Tasks

What is the relationship between an objective and a target?

Apart from the definitions provided by ISO 14001, most companies do not assign hard and fast meanings to the words 'objectives', 'targets' and 'goals', treating them all as largely interchangeable. If you are pursuing a formal EMS standard, it is important that you adopt the meanings given in the standard, no matter what the previous use of the words might have been within your company culture. In this book, we follow the lead of ISO 14001, and use 'objective' where we mean an overall goal, and 'target' where we mean a series of stepping stones towards that goal. In either case, and to aid management decisions, both elements should be measurable wherever possible.

It is also worth considering how you express objectives to the very different audiences in your stakeholder group. There is a world of difference between 'We are committed to improve local air quality' and 'We aim to reduce our emissions of carbon dioxide by 5 per cent over two years from a 2006 baseline'. The first could well be construed as an objective, and would certainly make sense to members of the public. However, as a target for the workforce, the second is far more useful in terms of what operational controls need to be used in order to achieve it.

It is also important to remember that objectives and targets are not static, even when they have been agreed at the beginning of a management cycle. Each objective may have a series of targets which relates not only directly to that specific objective, but may also contribute to or interact with other objectives and targets. For example, as a waste stream is reduced through greater process control or better management of resources, a potential revenue stream from selling on the 'waste' for recycling or further processing may also dwindle. The interaction of objectives and targets will require ongoing scrutiny by senior management as part of the review process (see Chapter 14).

Are there different types of objectives in an EMS?

The short answer is 'Yes', and such a variety is to be welcomed, because it can work to the benefit of the organization. Most managers talk of objectives as if they automatically implied an improvement; after all, why would anyone have an objective to get worse? Within the confines of an EMS, however, it is also possible to set objectives that either maintain legal compliance, or that research a particular issue further; objectives which, for convenience, we label 'Maintenance' and 'Research and development'.

When there is not enough expertise on a particular environmental impact within an organization, any manager would be able to see that it would make sense to undertake a specific review of the impact and what gives rise to it within the company. Thus, an objective that states that a review will take place, given further definition by a reporting back date, will be perfectly acceptable as an objective. Such a research programme may include further monitoring work, new measurements being taken and further data compiled.

In a similar manner, it may be obvious to the organization that a particular impact cannot be improved upon for a variety of reasons, especially if it requires the purchase of new technology that is not economically feasible at the time. If the impact is also being managed in such a way as to conform to legal requirements, there may be a strong case to create an objective of 'maintenance' or 'sustained compliance' in which the impact is carefully managed and maintained at current acceptable levels. This may still require some improvements in management techniques, risk assessment and operational controls, however, and should not be considered an easy option. There is still a requirement for the managers concerned to keep the circumstances surrounding the impact under constant review. What would happen, for example, if the law were to change and further measures were found an absolute necessity? What is the likelihood of such a change? Are the emergency procedures adequate to the continued task of containing such an impact under abnormal circumstances? All these questions require active attention even though an 'Improvement' objective has not been set.

If you are pursuing an EMS against a formal management system standard, the commitment to continual improvement means that not every objective can be a 'Maintenance' or 'Research and development' objective. However, a balanced approach, which takes into account the ability of the organization to respond to the seriousness of the impact will often find acceptance with the external assessors. Box 7.1 below gives an example of the three different types of objectives as they might be applied to the same environmental issue.

Box 7.1 Types of objectives in an environmental management system

Issue: Emissions of volatile organic compounds

Objective type	Objective
Improvement	'BIG Inc will reduce its annual emissions of volatile organic compounds by 25% over a two year period.'
Maintenance	'BIG Inc will comply with all current and future regulations concerning VOC emissions, monitor both point and fugitive emission levels continually, maintain management procedures for the use and disposal of VOC generating solvent materials and will review abatement technology options on an annual basis.'
Research and development	'BIG Inc will conduct a company wide review on their current levels of solvent use, and research potential non-VOC producing replacement substances. The company will also research alternative processes and procedures with a view to reducing VOC emissions. This review will be complete by (date) and this objective will be revised in the light of the findings.'

Within the context of ISO 14001 and EMAS, the policy requirements indicate a specific link between the policy and the objectives and targets, in that the first has to provide a framework for the second. In essence, this means that there should be no mismatch between the commitments contained in the policy (which will set out the stated EMS scope) and those in the objectives. Therefore, in practice, auditors (both internal and external) of the EMS would be looking for a balance of types in the objectives; a balance that favoured improvement over the other two types, and certainly a balance that stayed within the policy commitment to improve continually, prevent pollution and obey the relevant legislation or other requirements.

How is an objective made effective (how do I know I have set the right one)?

The mnemonic SMART may help here, especially if you use it as a template, and apply it to your final objectives and targets.

Specific: The broader the expression of an objective, the more complex the series of targets needed to address it. 'Being more environmentally friendly' is too vague an environmental objective for organizations which would certainly not allow 'Being more profitable' to be adopted as a financial performance target. Ask yourself the questions 'How will I know that the objective has been achieved?' and 'What will it look like when we get there?' to help focus on the important performance factors.

Measurable: The cliché that something cannot be managed because it isn't being measured became a cliché because it contains a fundamental truth of management. It also flags up the importance of choosing the right measurement (see 'Can environmental performance indicators help to set targets?' below) and where to take it, ensuring that it is relevant to the reduction or the improvement you are seeking. Specific substances, states or activities will provide the vital feedback on progress, so choose them with care.

Achievable: This word is at the heart of the balancing act involved in objective setting. Give people 'stretch' goals and they will surprise you with their ingenuity and resourcefulness, but put the goals too far out of their reach and people simply lose interest in being constantly judged as a failure. If you think a 5 per cent reduction is possible in a particular impact, aim for 10 per cent or even 20 per cent. If you then reach that target, think seriously about doubling it for the next year, and so on. It is all a matter of judgement, and environmental management is not an exact science, so expect a few surprises in the first year!

Realistic: This is the other side of the scales in which to weigh your objectives and, at first sight, it seems to be a principle in direct opposition to 'Achievable'. In reality, the truth lies somewhere in between, and can be discovered only if your analysis of the impact is matched to your setting of the objectives. In the first years of an immature EMS, there is no replacement for close monitoring of objectives and targets in order to see what progress is being made and if not, why not?

Time-related: All of the previous principles can be followed but if the time element is missing or in some way inappropriate, then achievement of an objective can be seriously hampered. Try to secure as many objectives as possible to give results in time for the regular management review meetings. This is the place where an overview can be most helpful, especially attracting contributions from those managers with a different perspective.

How are objectives and targets set?

There is no hard and fast rule in terms of setting objectives, mainly because no two organizations have the same combination of impacts, circumstances and business priorities. There are some useful guidelines that can be followed and will help to ensure that the objectives you set work within the context of your own organization.

PLANNING

Your environmental policy will probably have committed you to obey the law, prevent pollution and continually improve. If you are pursuing ISO 14001 and/or EMAS, it should definitely have done this. Your IER will have provided you with data on your significant environmental aspects and impacts, including any problems you have had in the past and potential problems in the future. It will also have provided you with an understanding of how the law applies to your organization and how that law will be developing in the short, medium and long term. You already have a series of business priorities, and possibly a business plan for the next five years. All these elements provide the information that you will need to balance in order to write the objectives and then set associated targets.

The key to the beginning of the process is the risk assessment element of the initial review. A simplified version of establishing this baseline will have been done for a phased implementation scheme, so the same approach can help here. The risk assessment will have helped to identify your environmental impacts in such a way that it should be relatively easy to group them into four headings:

1   High frequency/High impact;
2   High frequency/Low impact;
3   Low frequency/High impact;
4   Low frequency/Low impact.

This is not a particularly fine distinction at this early stage, especially when comparing very different impacts and circumstances. There will also be an element of subjectivity and personal judgement about the process at this point, but using a matrix and scoring system (see Chapter 4) is all that is required in simpler organizations. However sophisticated or simple your comparative techniques, ultimately, these four groupings will help you to prioritize your environmental objectives. Let's take a closer look at each sub-division.

High frequency/High impact: These are the impacts that should be at the top of your action list. A large adverse impact on the environment can hardly be allowed to continue unchecked, especially if your policy requires you to prevent pollution and stay inside the law. As a minimum, these impacts should be brought down to within legal limits, if they are not already. Where there is no legal requirement and the impact is still in this category, it remains the most significant type of impact relative to your operations and, as such, should be improved upon. Objectives have to be relevant as well as achievable. You will need to make this group your first priority for improvement objectives and action.

High frequency/Low impact: Under this heading are adverse impacts that apparently happen quite often but have a reduced impact on the environment. Care still needs to be taken here as the impact which is measured may be judged small with each individual occurrence, but may belie a much larger cumulative impact over time. If this is the case, the impact belongs more accurately in the 'High frequency/High impact' group. If this is not the case, look more closely at the cause of the impact itself, particularly at the time cycle of the occurrence, which will give you a clue as to which part of your overall process the impact is tied. The vast majority of impacts of this type are caused by breakdowns of process control, where a seemingly minor action as part of a larger process is repeated at frequent intervals. The solution can often turn out to be a relatively simple change in a work procedure, requiring very little in the way of extra resources to remove the impact altogether. Such solutions and objectives will feature prominently because of their cost effective nature: little effort for large gains.

Low frequency/High impact: Again, it is important to understand the basic cause of the impact in the first place. You may find that the circumstances that give rise to this type of impact are most closely allied to emergencies of some sort or a response to an accident or human error. This means that on a day-to-day basis, nothing will appear particularly problematic; accidents that are waiting to happen do not have a habit of making themselves obvious in advance. Blocked drains, leaky pipes, delayed maintenance, lack of emergency drills; all these factors can contribute to such impacts-in-waiting. As a result, although these impacts have great potentiality, the management effort and resources required to control them may not amount to very much. Again, though it is difficult to measure the cost effectiveness of good risk management with objectives in this area, no news is good news. If the cause is not linked to emergency situations, analysing the time base of the occurrences will help to define the cause and link it to specific activities on site. This in turn may well indicate that the impact needs to be re-prioritized as high.

Low frequency/Low impact: These can afford to be at the bottom of the list of the priorities when it comes to setting objectives. As an example of this type of impact, it may well be caused by a process that has an inherent characteristic that can only be solved through new technology, which may not be cost effective compared to the improved performance. On the other hand, be aware that in this area it is possible for an organization to have the perception that the impact is 'low', but that another group of stakeholders may consider it 'high'. Noise and odours often fall into this category, especially with members of the local community surrounding a works. Risk assessment by a company is not always the same as the risk perception by the stakeholders and it may be worth taking action in these areas if your stakeholder feedback indicates that there is a problem.

These headings provide a useful framework to help sort out some of the conflicting demands when it comes to setting priorities for action. The prime motivators for new EMSs will be to stay inside the law and prevent pollution, which in many minds is the same thing. If all your objectives reflect only these two aspects, however, you may only be managing your pollution and not your environmental impacts. This is where continual improvement comes in, which would indicate that the high impacts need tackling first. Business priorities will probably benefit more from areas where there can be significant financial savings and reduced payback times. These are more commonly found in the areas of energy costs and waste disposal costs at first, but using reduced environmental impacts as an indicator of greater process efficiencies will make financial benefits more obvious to senior management in the longer term.

PLANNING

Can environmental performance indicators help to set targets?

Absolutely. EPIs are simply areas of measurement that can supply information on the amount of progress made towards an overall objective. They are an integral part of the early stages in a phased implementation scheme, where the emphasis is on EPE as a way of managing impacts prior to the development of formal systems.

Having set an objective, thinking about which units of measurement need to be applied will help focus the mind on whether the objective is worth having, as well as the way in which an action plan can be formulated to deliver it. According to the international standard ISO 14031, it is useful to consider EPIs as falling into three broad types:

1   management performance indicators (MPI);
2   operational performance indicators (OPI);
3   environmental condition indicators (ECI).

If you are attempting to measure your progress towards an environmental management objective, it is a good idea to use two or possibly three different types, in order to 'triangulate' your current position from a number of different perspectives.

Management Performance Indicators: As their title implies, these look at the actions taken within the management function in order to realize the objective. They may measure such activities as policy formulation, team meetings, the drafting of action plans and so on; indeed any action in the management area that facilitates movement towards the objective. Experienced managers can probably already see that relying on these types of indicators alone will only measure the intention of the organization, and not much else. For example, if the objective is to reduce harmful air emissions from a paint spraying operation, the environmental management indicators for the series of targets might be the existence of:

- air emissions policy;
- air emissions objectives and targets;
- air emissions action plan;
- paint spray booth work procedures;
- training needs analysis carried out/training programme designed.

Operational Performance Indicators: These, on the other hand, are what most people think of when indicators are mentioned in the field of environmental management. These are appropriate measurements of the actual amounts of emissions, discharges and other quantified environmental impacts that are caused by the organization's activities, services and products. (If preparing for an EMAS public statement, the figures included will all fall into this category.) They measure the implementation of the management intent, and whether policies and procedures are being adhered to by the rest of the organization. Again, taking the example of the paint spraying operation, the OPIs might be:

- existence of any abatement equipment (and associated maintenance programme);

- existence of air monitoring equipment (plus calibration programme, monitoring programme,
- record keeping, auditing of results, checking and corrective actions, etc);
- performance/output of abatement equipment;
- adherence of sprayers to work procedures;
- training of sprayers in work procedures.

Environmental condition indicators: These indicators apply directly to the state of the environment in the locality, region or nationally. Obviously, attempting to correlate your actions to a national ECI can be difficult, so it is more effective to work with your local authority to see what ECIs they are already using and how you might perhaps quantify your own contribution within their framework. These are the only sets of measures that apply directly to the effectiveness of both managerial and operational indicators.

For a paint spraying operation, the ECIs could be linked to Local Authority Air Quality Plans where they exist, and the specific measurements of volatile organic compounds could be part of that framework. As each authority may increasingly have to provide national government with a planned series of targets for air quality improvement (a growing practice in European countries), an individual organization's contribution could be used as part of the figures produced for that plan. Note that the output of the paint spraying abatement equipment is not an indicator of the condition of the local environment but an OPI.

What other elements of the EMS are affected?

Policy
(ISO 14001 Clause 4.2 /EMAS Annex I-A.2)

As mentioned above, the policy element of an EMS helps to provide the framework for the subsequent objectives and targets. As a guide, if there are more objectives than are mentioned in the policy, this may not be a problem; it means that you are undertaking a broader programme of improvement than you are stating to the public, which may be your choice. On the other hand, having more policy points than objectives can be a problem; it means that you have stated publicly that you are addressing a problem, yet haven't set any objectives to deliver against this statement. This would almost certainly be regarded as a non-conformity by visiting external assessors.

Managers also often ask whether it is necessary to include the objectives in the policy statement itself in order to meet the requirements for EMAS and ISO 14001. The answer is simply 'No'. There is nothing wrong with including them, but managers should bear in mind that as objectives change over time, their inclusion would probably lead to redrafting of the policy statement more frequently than if they were not included.

Environmental management Programme(s)
(ISO 14001 Clause 4.3.3 /EMAS Annex I-A.3.4)

A management programme or programmes support objectives and targets by providing for their delivery. They are an integral part of the standards requirements by being included in the same clause and can be broadly characterized by the phrase 'who does what, by when'. However, the most important link between the objectives and the management programmes is that there should be consistency between the two. It is remarkably easy to overlook a particular objective and not allow for it in the drawing up of responsibilities and subsequent action plans. Again, external assessors finding a gap between the two elements will in all probability consider their findings a non-conformity against the standards.

Monitoring and measurement
(ISO 14001 Clause 4.5.1 /EMAS Annex I-A.5.1)

Objectives require measurement, which in turn requires the monitoring of emissions and impacts, which in turn requires records. All of this presupposes that the most appropriate measure has been selected for a particular objective or target. If you are unsure, check any legislation for specific monitoring requirements as a minimum. Where this may not prove to be enough, try using your local regulator for further guidance. Again, consistency of approach and the overall facilitation of delivery is important here. Make sure that there are no sins of omission – to avoid sins of emission.

 Narratives

Smallco

The EMS is now up and running. The hard work to carry out an IER, establish legal and other requirements baseline and policy work has been followed through by establishing focused objectives and targets for the EMS. Unfortunately, legal compliance issues aside, most of these are for short-term payback; for example, energy savings and waste minimization. The Ops Manager is slightly concerned that once these quick fix solutions have been implemented, the EMS is in danger of fizzling out. Perhaps the strong focus on quick fixes back at the commitment phase might have been tempered with some examples of medium- and long-term benefits to be gained from environmental management.

The Ops Manager decides to work from these positives and jots down some notes for the environmental slot at the next management meeting:

- present projections of short-term gains we are on course to make, and the business benefits that accrue;
- establish relevant milestones to measure progress;
- suggest first steps for key issues that are medium- and long-term objectives for the EMS and flag up potential benefits;
- suggest some of the returns from the short-term gains – perhaps 10 or 15 per cent – could be invested in work on longer term objectives with identified business need;
- illustrate these potential returns to us with some examples from other businesses – the major customer has agreed to provide some information here;
- select three or four areas for action now and indicate that others will be put on hold until next year. The pilot work on these three or four will

provide important learning for ranking and working on the rest. Some may even be shelved for now. Reinforce that it doesn't make sense to try to do everything at once and that we do need to do things which make business sense.

BIG Inc

In some areas, the project manager discovers that monitoring progress towards the objectives is reasonably easy (ie better energy management within company buildings), but in many others, defining meaningful indicators is harder. The usual practice within BIG Inc is for the key performance indicators to be part of the client brief, but it rapidly becomes clear this will not be enough for the EMS objectives. The reasons for this were again linked to the indirect nature of the impacts the company is trying to influence, the nature of the organization itself (lots of home-based workers) and the length of the supply and client chains upstream and downstream of their operations.

Looking for help and guidance on the matter, the project manager turns to ISO 14031, and quickly realizes that a mixture of indicators is needed. Because of the nature of the objectives, MPIs are going to be plentiful and relatively easy to measure. That said, OPIs will be needed for those working away from headquarters, and included in the client briefs where appropriate so that some kind of progress can be measured. ECIs could only be used at first in conjunction with certain local objectives linked to energy and transport policy at the headquarters building. In the long term, the company identified that further work would be needed to develop relevant and usable ECIs for the consultancy and design operations; tough but not impossible.

Trade secrets

- Having objectives without targets is like drawing a map in the sand, then shovelling the sand into a bucket so you can read it later.
- Getting buy-in from managers in other functions works best if the objectives are not out of alignment with their own business, or even personal performance objectives. Make it work for all concerned.
- Use cost benefit analysis to help prioritize objectives, but not to decide upon the significance of an impact.
- Go for consistency. Make the policy match the objectives, match the targets, match the management programme, match the audit programme.
- If you're managing it, you're measuring it. If you're measuring it, make sure you're measuring the right thing at the right time.
- Use different types of measures (EPIs) to take different soundings at different points of your journey towards your objectives.

Things to think about

- As phased implementation schemes continue to develop, direct inspection of environmental contractual requirements may also grow in popularity. The structure of such schemes allows for the external inspection of the development of individual EMS elements, prior to the finalization of the whole system. It is only a small step to include customer-defined environmental requirements in the inspection process, especially as EPIs usually feature strongly at an early stage of the overall EMS implementation. They may even become a part of client-supplier (second party) audits.

- Given the complex and interrelated nature of ecological issues, ECIs are the most difficult indicators to identify and apply. It is no wonder that many organizations often settle for information from management and operational indicators alone, as for many executives these help to define the bounds of organizational responsibility. As the ethical discourse on the subject of corporate responsibility continues to evolve, however, such boundaries may have to be revisited.

8

Environmental Management Programmes:
Who Is Doing What by When?

Your EMP is the vital link between thought and action. It is the blueprint that brings to life what you plan to do as set out in your environmental policy statement and flows from your detailed objectives and targets. It achieves this by establishing a systematic implementation programme or series of interlinked programmes and prioritizes roles, responsibilities, processes, resources, timetables and is designed to provide a dynamic plan of campaign. A well-designed programme helps to foster commitment through the organization as it begins the implementation and ongoing management of an EMS in a logical and pragmatic way.

✓ ISO/EMAS quick check

Area of EMS	ISO 14001	EMAS
Objectives, targets and programme(s)	4.3.3	Annex I-A.3.4
Resources, roles, responsibility and authority	4.4.1	Annex I-A.4.1
Monitoring and measurement	4.5.1	Annex I-A.3.4

☐ Chapter executive summary

Move to the next chapter when you can answer all of the following questions in ways which make sense to you and your organization:

What resources are required: human, physical and financial?
What is project management and programme control?
How do I develop the programme(s)?
How do I measure and monitor progress?

⟰ You are here

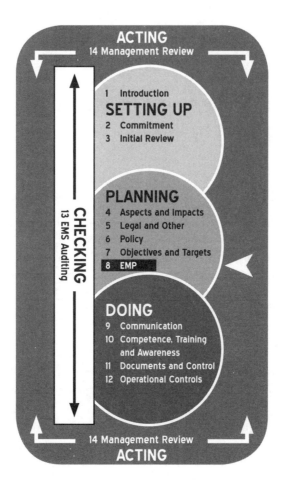

⛏ Toolkit requirements

To work through this chapter you will need the following knowledge and skills:

Knowledge

- thorough understanding of your business issues;
- clear and appropriate objectives and targets;
- actual resource allocation;
- performance indicators.

Skills

- developing action plans and SMART objectives;
- liaison skills;
- general planning skills;
- measurement and monitoring skills.

Tasks

What resources are required: Human, physical and financial?

The importance of securing appropriate resources for your EMP cannot be underestimated. In both ISO 14001 and EMAS the identification of appropriate resources is clearly identified as an essential element for the implementation and running of an EMS. The allocation of adequate resources is vital to sustain the EMS through its entire life cycle. An effective EMS will track benefits as well as costs; demonstrating the value of return from investment in the EMS will in turn help to secure and sustain the resources required.

Obtaining resources for new work is a major challenge in any organization. Often, the resource base is actually shrinking as organizations are required to do more for less in an increasingly competitive business climate. Also, the benefits of improved environmental management may not flow directly back to the bottom line immediately and investment funding will undoubtedly be required. These are major challenges for the environmental manager charged with the responsibility for bringing an EMS to life.

Clearly, there are some key areas that must be resourced. Not only that, but the organization will need contingency plans to ensure that these key areas have resources that are uninterrupted (ie due to production fluctuations, illness, etc).The most obvious is perhaps to fund work to ensure that the legislative demands, both environmental and health and safety, are actually met. These environmental liabilities bring clear obligations that must also be met. In addition there are also many low cost initiatives that will yield significant fiscal and environmental benefits. Waste minimization and energy saving are two obvious examples. In the particular situation of your own business there may be other less obvious but equally significant opportunities to show short-term fiscal gains which can release resources for funding longer term initiatives. As we discussed in Chapter 2, securing commitment from senior management is a cornerstone of an effective EMS. Securing adequate resources is a clear signal from senior management that this commitment is tangible.

All business decisions have resource consequences. Most managers will have had direct experience in developing the business case for winning resources. The skill employed to understand and determine these arguments applies equally to developing the business case for environmental management. Without understanding and determining the resource requirements for environmental management, and being able to communicate this information to other managers in the organization then you cannot hope to control your EMS with any confidence.

As with other management decisions, an actual decision – in this case on programme design – is reached by analysing a number of realistic alternatives. Should you buy in legal expertise to help determine applicable environmental legislation? How should you develop internal skills to implement changes in procedures? How can you best bring customers and suppliers up to speed on what your new expectations are of them? The list is endless. The important skill is to be able to evaluate the alternatives – and there will always be the 'alternative' of doing nothing – against your own objectives for the EMS and those of the business itself.

The underpinning that will advise the resource allocation process in your EMS can be summarized in five steps:

Step 1   Set the objectives (see Chapter 7).
Step 2   Define the issues (see Chapter 4).
Step 3   Identify alternative options for action.
Step 4   Evaluate these alternative options against your EMS objectives.
Step 5   Select the best option and implement it.

For example, the decision on whether you should buy in legal expertise to help determine applicable environmental legislation and keep this up to date will involve:

Step 1   Set the objectives: determine applicable and future legislation.
Step 2   Define the issues: how best to do this?
Step 3   Identify alternative options:
      • extend existing in-house expertise;
      • develop new in-house expertise;
      • buy in support; internal/external;
      • subscribe to external information provider and write procedures for their use;
      • do nothing.
Step 4   Evaluate: quantify each option in terms of time, cost benefit and output in terms of meeting EMS objective.
Step 5   Select: what is the best option in terms of both meeting the objective and resource utilization?

Clearly each one of your objectives set in Chapter 7 and re-set on an ongoing basis will require finding answers to the numerous questions raised by following the five steps suggested above. Following such a procedure provides a rigorous analysis of options and will advise the selection process. It also provides hard objective evidence to allow reasoned argument in putting the case for adequate resources to bring the EMS to life. Of course, in reality it will not be possible or desirable to do everything at once. Overlaying this resource decision making process with your earlier work on aspects, impacts (Chapter 4) and objectives and targets (Chapter 7) will ensure that the priorities you identify fulfil the spirit of an EMS which is to improve environmental performance in ways which are compatible with the business itself.

What is project management and programme control?

Guided by the environmental manager it is now the time for the management team to roll up its collective sleeves and get on with making things happen. The environmental programme(s) forms the framework for implementation and running the EMS and will include strategies for measuring and improving environmental performance. The programme(s) can be stand alone and achieve some limited success. It is, however, likely to be much more effective if it is integrated with the organization's other business and strategic plans. The EMP must be a dynamic creature, able to respond to change in the EMS itself and also to reflect any changes in wider organizational objectives and targets.

An EMP is the antidote to dealing with environmental issues as they arise. As we discuss throughout this book, being proactive puts you and your organization in control. Responding as the need arises suggests a reactive approach, dealing with issues after the event and addressing symptoms, not developing cures. The formal EMS standards, not surprisingly, require corrective action to ameliorate the effects of environmental incidents. They also require preventive, and therefore proactive, strategies to be implemented to ensure the original problem does not recur and so that sustained improvements in the organization's environmental performance actually take place. ISO 14001 and EMAS aside, the underlying rationale for being proactive runs much deeper. A comprehensive management programme which identifies specific actions in a prioritized order will provide superior assurance that all significant environmental aspects will be dealt with in an appropriate way.

Often, inadequate resources and lack of support are specified as major constraints which prevent an effective EMS. Support for an EMS does and will require a change in attitude and corporate culture. Leadership from the environmental management team and its demonstration through a clear, effective and well-planned EMP will help foster commitment, keep senior management on board and facilitate change.

How do I develop the programme?

Rudyard Kipling, in a short poem to provide advice to young journalists, provided a splendid model for developing an EMP. His poem presents the reader with six straightforward questions, represented as servants' names. Only when the answers are in place to all six questions is the EMS programme watertight. Here is his poem:

> I keep six honest serving men
> (They taught me all I knew)
> Their names are What and Why and When
> And How and Where and Who
>
> (Rudyard Kipling)

So for the EMS the questions that emerge are:

Question . . .	Response . . .
What do we need to do?	Your work in IER, aspects and impacts, legal and other requirements will have revealed the answers for your organization.
Why do we need to do it?	Your analysis of stakeholders and business drivers will have clarified the answers to the why question.
When do we need to do it?	Your answers to what and why will signal when you need to do it.
How will it be done?	Setting SMART objectives and targets to bring your environmental policy to life will signal how.

PLANNING

Where will it be done	Perhaps the easiest question as the nature of the 'what' answers will tell you where the focus of effort is required.
Who will do it?	As we have discussed in this chapter, well-defined roles and responsibilities are of paramount importance. If no one is actually responsible then nothing will actually happen.

What might the programme look like?

There are numerous ways to set out a management programme. Several are suggested here and are drawn from our extensive experience as ways that work in practice. Whichever way you actually choose, try and follow the seven steps set out below. If you can adequately answer and address the questions and issues raised in each step then your EMP will begin to settle into place.

Step 1   Information:
- what do you already know?
- who has the information?
- what form is it in?

Step 2   Involve others:
- who are the key staff?
- arrange to meet key staff;
- how can you get them on board?

Step 3   Set targets. Get SMART and set targets that are:
- **S**pecific: what will you do?
- **M**easurable: how will you know you've done it?
- **A**chievable: can it be done?
- **R**ealistic: does it make sense?
- **T**ime-related: when can you do it?

Step 4   Plan:
- agree a plan together;
- write it down;
- who will do what?
- resources required;
- set a timetable.

Step 5   Commitment and leadership: if they are not already, get senior management on the side of environmental management.

Step 6   Action: get on with it! (See 'Action plans' below.)

Step 7   Review your progress:
- keep a check on what is happening;
- does it match your plan and expectations?

Action plans

Action plans form the basis of Step 6 above. They should identify the targets you have set and provide points of reference to enable you to measure and monitor progress towards achieving them. To build an action plan you need to consider:

- the broad areas in which you want to achieve an environmental result, as established in Chapters 4–6;
- what are your actual objectives and targets (the key outcomes from Chapter 7);
- the timetable for starting and finishing each step in the programme; and
- who needs to be involved; internally (eg colleagues and internal expertise) and externally (eg regulators and local support networks).

Table 8.1 illustrates some key elements used to form a programme action plan. However, as noted earlier, there is no right way to design your programme. You need to do what works for you in your organization. It makes good sense to investigate related documentation in your organization which might help provide a framework for the development of your environmental action plans. It is often useful to sub-divide each item in the generic action plan into a series of smaller steps. This will enable the plan to be implemented in an interlinked series of more manageable chunks. An example of this structure is shown in Table 8.2.

Table 8.1 Example action plan

Result area	Output	Time-scale	Responsibility	Assistance required
Waste minimization	Reduce all waste by 20% and cost of disposal by 40%	End 200?	Env. Manager	Site Manager Line Managers Waste regulation bodies
Energy efficiency	Reduce energy usage by 10%	May 200?	Site manager with Env. Manager	Advice agencies Energy suppliers Technicians
Effluent treatment	Reduce levels of chemical usage Reuse?	Jan 200?	Env. Manager	Site Manager Consultants Professional bodies
Transport	Reduce number of shipments	End 200?	Transport Manager with Env. Manager	Customers Designers Transport bodies
	Increase fuel efficiency of car and commercial fleet	Sept 200?	Transport Manager	Drivers Advanced motoring school

How do I measure and monitor progress?

Running an EMP is a dynamic activity. Achieving each target and objective may be viewed as a multi-strand project, so learning from experience and applying this understanding to the process is a vital part of effective project management. To be effective, this needs to be done through regular measurement and monitoring of performance. ISO 14001 and EMAS both

Table 8.2 A step-by-step environmental action plan

Objective(s): Waste minimization • reduce all waste by 20% • reduce disposal costs by 40% **Resources needed:**			Task started Date	
			Deadline Date	
			Time allocated 10 weeks	
Staff time Investment – to be calculated New procedures			Priority importance: score out of 5; speed; score out of 5 and multiply together **Total:**	
			Task completed	
Key steps (action breakdown)	**Who involved?**	**Challenges**	**By**	**Completed**
1 Waste audit (use IER methodology)	Environmental Manager	Collecting numeric data	Week 4	
2 Define objectives and targets for waste minimization	Environmental Manager Line Manager	Reaching agreement	Week 6	
3 Liaise with waste regulators for advice	Environmental Manager Site Manager	Prioritizing ideas	Week 4	
4 Develop draft action plan for agreed objectives and targets	Environmental Manager	Writing this in a user-friendly way	Week 7	
5 Report (to senior management)	Environmental Manager Site Manager	Achieving buy-in	Week 8	
6 Brief (Line Managers and staff)	Environmental Manager Line Manager	Achieving buy-in	Week 9	
7 Develop infrastructure to support action	Environmental Manager Line Manager	Ensuring focused investment	Week 10	
8 Monitor and refocus action	Environmental Manager	Time	ongoing	

Notes:

emphasize the importance of continual improvement. This involves the ongoing refinement of your EMS, continually improving the management of significant environmental aspects of the business in order to improve environmental performance and reduce pollution.

To ensure your EMS and the programmes to implement it connect seamlessly, two competences are needed: the ability to both measure and monitor performance. These competences apply to all aspects of EMS management.

1   If you can't measure it then you can't manage it.
2   You get what you inspect, not what you expect.
3   If you want to know what someone thinks is really important, you should look and see what he or she spends his or her time on.

(Sir Anthony Cleaver, Tomorrows Company, RSA Lecture 1995)

Ask yourself three questions about your installation plans:

•   Did the programme achieve the specified objectives?
•   Did the programme actually improve environmental performance?
•   Was the plan worthwhile?

In this final section of the chapter we will look at finding answers to these principal questions. Firstly, it is worth reminding ourselves of what measurement and monitoring mean in the context of environmental management.

Measurement

Measurement is an essential component of both maintaining and improving environmental performance. An unblemished record on environmental issues, such as compliance with regulations or zero pollution incidents, is not necessarily a cast iron guarantee that environmental risks are being controlled. Even if an incident at your site may be highly unlikely because of the particular nature of your operations, the consequences of a single accident may be devastating. Effective implementation of environmental management demands effective measurement of the impacts. Having control of the data is a key task. The data will signal the areas where potential risks exist and the likelihood of these risks becoming hazards. The data provide the baseline against which environmental progress can be measured.

Environmental performance can be measured in two ways:

1   Negative performance indicators such as accidents, incidents, direct interventions from regulators, staff absence due to environmental issues, damaged equipment, down time and lost production while having to take remedial action. These can all be quantified for the organization.

2   Auditing and sampling which, alongside the IER work in Chapter 3, provide ongoing analysis of the effectiveness of the EMS.

There is a number of increasingly sophisticated EPIs that can be developed to measure performance. The guidance standard *ISO 14031, Environmental Performance Evaluation* provides excellent advice on performance indicators using three indicator groupings. As we pointed out in Chapter 7, the first two relate directly to organizational performance and the third provides the direct link with improved environmental performance.

The establishment of relevant EPIs provides an important mechanism to measure organizational performance across a range of meaningful criteria. However, measurement alone will not result in effective environmental programme management. It needs to be linked to monitoring.

Monitoring

Like measuring, monitoring signals a commitment to environmental management. To be effective, monitoring needs to reflect the realities of managing an organization. Two types of monitoring, active and reactive, will ensure that the day-to-day realities of organizational operations are addressed by the environmental management programme implementation:

1   *Active monitoring* = preventive action. Active monitoring is the essential feedback on performance against your objectives and targets. It involves checking compliance with performance standards – your own or those imposed by regulations – and the achievement of the specific objectives which you have set to manage the significant environmental aspects of the business. Its key purpose is to demonstrate success.

2   *Reactive monitoring* = corrective action. Reactive monitoring manages deficient environmental performance; for example, accidents (to people and the environment) incidents, hazards and direct non-compliance with regulations.

These competences – to measure and monitor environmental performance effectively – are promoted through training and communication, the goals of which are to create the organizational culture which emphasizes a diligent approach by all staff to environmental management. However they are actually presented in the organization, your goal must be to ensure they can be accurately interpreted by management to allow the required corrective actions.

Once again, the monitoring of the programme needs to address constantly the eternal triangle of time, cost and quality. Your work to measure and monitor the ongoing implementation of the EMS should provide regular and accurate feedback loops (see Figure 8.1) so the project itself is constantly being fine tuned where it needs to be. The purpose of this work is to control the installation project and this should seek to give those staff involved positive opportunities to improve the EMS constantly. The ultimate goal of the project is to deliver on the objectives and targets you and your colleagues have set to reduce the environmental impacts of the business.

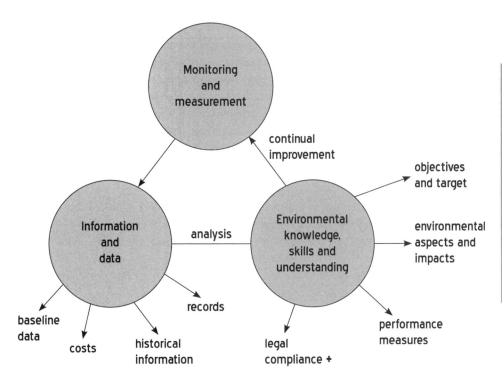

Figure 8.1 Simplified map of an EMS

Narratives

Smallco

Smallco's phased approach based on BS 8555 and the associated EPE has worked well and focused management attention on areas that bring business and environmental benefits. Agreed objectives and targets have now been set and work is well under way to establish and operate programmes of work to action them. The support of key customers has been critical in achieving significant progress to date and has strengthened working relationships. The EMS is beginning to take shape. However, despite the best efforts of the Ops Manager, departments are doing what they think makes best sense rather than what they have been requested or even agreed to do. Several paper recycling bins have materialized in the offices, and freight transport association 'Save energy/drive efficiently' posters have appeared in the dispatch area. The challenge for the Ops Manager is not to stifle these individual initiatives but to keep programmes of work on track with priority objectives.

The Ops Manager takes the time to talk with several staff at various levels in the business. What is apparent is that they are not clear about the emerging big picture and what it means to them. They know something serious is underway and want to be seen to be doing something. The Ops Manager talks with his management colleagues and they agree to let him do a short presentation at the next operational meeting(s). The presentation will acknowledge the existing efforts being made and make five key points: *What* the emerging EMS is about; *What* it means for Smallco and staff; *How* the objectives have been established; *What* they will deliver in terms of improved environmental performance; and *How* staff can help in achieving them through programmes of work. The Ops Manager also agrees to provide quarterly updates on progress and outcomes of this new work.

Smallco have now completed Phase 3 of their phased implementation and are ready to formalize their EMS as it rolls out into implementation and operation.

BIG Inc

Though the earlier LCA workshop helped to raise awareness of the possibilities for BIG Inc's EMS among line managers, when it comes to being confronted with the EMP, objectives and changes in work practice, the line managers do not respond positively. Even though the company is commercially successful, managers already suffer from role ambiguity and feel overloaded. Compared to the tough financial targets that they are being driven to achieve, the relatively 'delicate' nature of the EMS objectives look to them to be vague and inconclusive. Though the business culture of the company may have identified its core of values, it becomes obvious that some values (financial) are supported more than others (environmental).

The project manager takes the bold step of holding an open forum so that managers can voice their concerns. At first the feedback is overwhelmingly negative, though it becomes obvious that the root cause of the problem is the fact that the EMS is not seen to support the overall business objectives. Rallying somewhat, the project manager calls on the Marketing Director to give an impromptu presentation on why the board see environmental issues as core business issues. This appears to have a positive effect and the presentation is reinforced some days later by a personally addressed letter from the Chief Executive on the same subject, emphasizing that the EMS is important enough for individual performance indicators from the programme to be included as part of the staff annual review system.

Trade secrets

- Start at the end not the beginning. Decide where you want to go then walk backwards to the start so you can see all the steps along the way. All effective project management happens this way.
- Remember, perception is reality. People do what they think is best. If you don't help them to focus on what is best for the project we call the EMS they'll do something else.
- Know your culture. Don't set up a project which would be perfect for a supermarket if you're working in the chemical industry. The EMS project must be relevant to your setting and staff.
- Establish clear measures for success. Rewarding progress is 20 times more effective than punishing failure.

- Lead by example. It sounds obvious but is often overlooked. People respond to what they see going on, not what they are told to do.
- If you have communicated effectively with staff, trust them to get on with the job.

⁇ Things to think about

- Techniques relating to EPE will become increasingly sophisticated. Although ISO 14004 is a useful start point, trade associations and professional bodies will be working to make the techniques relevant to their own sectors. Performance indicators provide a mechanism to measure progress but as the indicators themselves develop, some progress may turn out to be not as impressive as was first thought.
- EPE was originally intended to be an alternative to using an EMS, not just an adjunct. Expect to see the rise of contractually related environmental performance requirements as a result.

PLANNING

Doing

9

Communication: Who Talks to Whom?

Communication is the glue that holds the EMS together and is the process the organization uses to provide and obtain information. A key role for the environmental manager is deciding who – both internally and externally – needs to know what, when they need to know it and what they need to do as a result of this new knowledge and understanding. As ever, a systematic approach will deliver the best results and help focus on maintaining environmental performance. (*ISO 14063 Environmental management – Environmental communication – Guidelines and examples,* is an excellent aide-mémoire for developing such an approach). As we have said before, people who don't know vote no. For an effective EMS you need them to vote yes. Effective communication will ensure that this happens.

✓ ISO/EMAS quick check

Area of EMS	ISO 14001	EMAS
Communication	4.4.3	Annex I-A.4.3
Operational control	4.4.6	Annex I-A.4.6

⬛ Chapter executive summary

Move to the next chapter when you can answer all of the following questions in ways which make sense to you and your organization:

What are the steps to effective communication within your organization?
What is internal communication and why does the EMS need it?
How do I implement a communication programme?
What is our decision(s) regarding external communication issues?
How do I motivate people?

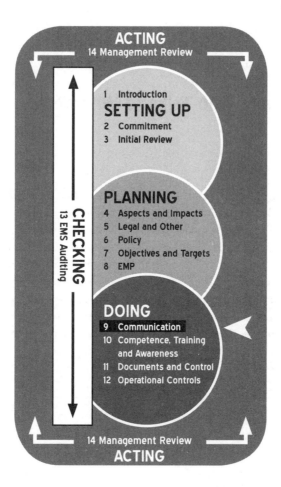

⤵ **You are here**

Toolkit requirements

To work through this chapter you will need the following knowledge and skills:

Knowledge

- key staff you will need to communicate with;
- understanding of the organizational culture;
- who are the stakeholders?
- what motivates people.

Skills

- listening skills;
- observational skills;
- motivation skills;
- translation skills – from jargon to business-focused language;
- written and oral communication skills.

Context

Communication plays a key role in making sure that things get done. It involves everyone whether they like it or not. Even 'not communicating' will have an effect on what happens in a business. Most people want to do a good day's work. If they are not totally clear about what is expected of them, they will fill in the blanks and take what action they believe is best (often called common sense!), which may not be what is actually best for the EMS. Clearly, an operating EMS requires many things to be done, not as single discordant activities, but as a harmonious symphony, with the environmental manager as the conductor!

Most of our communication skills are learnt in social rather than organizational situations. In social situations, communicating with peers or family, much is left out in terms of spoken communication and the gaps are filled by the recipient of the discourse. Experience tells us how best to fill the gaps and even if we get it wrong the outcome is not usually that serious.

This communication skills base may not always be best suited to the business situation where miscommunication or lack of any communication results in people filling in the gaps themselves, which might mean the difference between following a required EMS procedure or causing an unwanted impact on the environment. The driver of the lorry who discharged aluminium sulphate into the public water supply in Cornwall, UK, believed that he was following procedure, doing his job properly and doing the right thing. The outcome, however unintentional, was a serious contamination of the public water supply and the low level poisoning of many thousands of local people and their pets.

If all the many communications that need to take place to run an EMS are to make sense, then some kind of systematic approach to managing the communication process is essential. Not only do people need to listen, they need to develop an understanding that what you are saying to them is significant. Compare the following scenarios; which makes best sense and should lead to some positive action?

Scenario 1

As some of you probably already know we are now dealing with environment as well as everything else. It is important that you think about this and find out what the environmental aspects are of your job or work activities you are involved in. Remember that if you get it wrong then the regulators might close us down and that means major problems. Any questions?

Scenario 2

Some of you may have heard that we are beginning to develop an environmental management system (EMS) for the company. We are going to do it in ways which make sense for the business and that means the staff too. Part of today's session will be to identify some of the tools and techniques we can use to recognize some of the things we might be able to do to improve our management of environmental effects. I also want to put this into the wider context of our EMS and make the links with how your efforts will make a real difference.

When a person communicates with a colleague, they are conveying ideas and feelings: both essential components of effective teamwork and the key components of the communication process. A group of people must be able to share ideas and feelings with one another. Without this sharing they are only able to operate as individuals. If an EMS is to work, then in turn the communication process needs to be managed.

Communication is not just about telling people things. This suggests it is a simple one-way activity and assumes the recipient is listening and knows what you are talking about. Many of the ideas which environmental management brings to the workplace, and the changes in behaviour which will be required, are new and challenging. Everything that is done in an organization communicates something. Dirty and poorly maintained washrooms may say to staff 'we don't really care about your welfare here' and will result in the negative message being carried by staff to the workplace. Your staff will respond to what you are communicating, whether you have communicated it knowingly or not. The outcome may not be in tune with what you want to achieve.

Staff represent the greatest threat to any project and the greatest potential aid to its success. Without employees on your side then the best laid plans will fail. However, if staff are sold on the ideas in your plan then the project will almost run itself. Every aspect of staff communication must reflect your organization's commitment to work towards the highest standards. Remember, for the whole organization, environmental management and the drive towards continual improvement are a journey and not a destination. Making sure everyone has the same map to work from and tools they need are the keys to successful communication and training.

 Tasks

What are the steps to effective communication within your organization?
If your EMS is to work effectively then it is essential that you establish clear, concise, continuous and connected communications: the 4 'C's of communication. Think about some 'communication' you have received recently. Perhaps a professional telephone conversation or a letter or memo. Would it pass muster in the 4 'C's test?

- *Clear*, so that there can be no misinterpretation.
- *Concise*, so that people are not subjected to an after dinner speech each time you communicate.
- *Continuous*, not non-stop banter, but so that the communication process is ongoing and systematic and refreshes all parts of the organization.
- *Connected*, so that it makes sense to the receiver and connects with their own personal and professional lives *and* to the EMS itself so that they can find their own personal cross-references.

The cross cutting nature of an EMS means it knows no boundaries especially within the organization. This means as a manager with responsibility for all or part of the system there is work to do to ensure that the nature of this distinctive creation called an EMS is clearly understood by all who need to know it. In reality all staff within an organization and many stakeholders outside it will need to be engaged in your communication process so that the EMS can function effectively. There are five areas which will require attention to ensure efficient and effective communication:

1 Know your audience

Although this might sound obvious, you can probably think of several recent 'communications' which failed to address your needs and failed to make clear sense for you. Knowing your audience is *more* important than knowing what you are talking about. Listen to a good political orator. They are often thin on content and yet each member of an audience can find something concrete to identify with. This doesn't happen by accident and is the result of excellent research work by the politician's team as well as their own skills. People in functional units in the organization are often dedicated to their own objectives and targets and dismissive of externally imposed new and unwelcome changes. You might perceive the germinating EMS as a benefit, another group of staff might well see it as an unwelcome threat. So get to know your audience and communicate with them, not at them.

2 Make sure people do more than just talk about the issues

Assuming that your communication strategies bring people on board so that they buy into the EMS, there is a danger that they might stop at the intellectual level. Transport policy is a good example of this effect. At an individual level people feel strongly that public transport should be improved and there should be fewer cars on the road. This is a good objective with associated targets at a national level to manage a large environmental impact and reduce carbon dioxide emissions. It's a great idea but we don't put it into action by switching from cars to rail or bus. When you communicate all or part of your EMS make sure you do so in ways which will enable people to move from awareness to action. There is likely to be consensus on the fact that something needs to be done. You need to provide or better still develop concrete ideas which will deliver environmental improvements. As Sir John Harvey Jones, former CE of ICI and business guru, said 'It is easier to declare intent than to carry it out'.

3 Know the culture

This applies at a multitude of levels in the organization. The culture is like an invisible force which you need to get on the team! People will process and act on your communication messages in ways which fit the culture more than they fit your goals. Develop a good understanding of the broad corporate culture and the variants of this in different departments and match your communications to these realities rather than imposing your own. Remember, culture is based on how management actually acts not on how management says it will act.

DOING

4 Environment = emotions

If there is one area in business which realizes emotions more than any other, it is the environment. Emotions might include:

- *fear* of being found out to be in breach of relevant environmental legislation;
- *passion* and *enthusiasm* for fighting the 'green' corner and protecting the planet and our children's future well-being;
- *anger* at the degradation of the planet by 'fat cat' industrialists;
- *guilt* about working for a polluting industry; and
- *delight* that the organization is going to do something constructive to manage its environmental impacts.

An EMS which brings people on board will bring their emotions on board too. These emotions need to be seized in a positive way by providing opportunities for people to channel these energies into making a beneficial difference.

5 Jargon or professional language?

Environmental management brings with it the baggage of professional language and unprofessional language. The word 'green' is so overused and steeped in historical images of sandals and lentils that it has ceased to be a useful adjunct. Other jargon has pervaded environmental management: life cycle assessment, best practicable environmental option (BPEO) and Agenda 21 to name but three. The Glossary at the end of the book sets these and others out in more detail. Part of the job in communication is to demystify this professional language and help people make their own sense of it. Ideally, outside the EMS team, it should be avoided completely, especially in the first few years of EMS implementation. If not it can quickly become dismissed as simply jargon: that special foreign language to be rejected and ridiculed at every possible opportunity. The path to EMS installation becomes steeper still if people are listening for fun rather than acting on the content of the message. If at all possible avoid terms like 'green' or 'environmentally friendly' in your communications; they have come to mean almost everything to everyone and can undermine your professional strategy.

Internal communications

To communicate effectively you need to think about where you are now and where you want to get to. If someone is standing on your foot then effective communication will remove the source of the discomfort quickly. A message is transmitted, understood by the receiver, they take the action you want and a desired outcome is achieved. A communication programme for your EMS is not really very different – it just has more parts to it. Communication can be considered in four categories:

1 Informing

You might simply need to give staff the information required to change their actions, for example:

- Implementation of our EMS means we will be reducing consumption of raw materials per unit of product by up to 5 per cent over the next 12 months (a fact).
- I think we could save at least 12 per cent on transport costs if we reduced product weight and packaging (an interpretation of facts).

2 Instructing

Instructing is a much more directed communication, usually with a specific purpose in mind. For example:

- Now that we are returning (what were) waste raw materials to the supplier it is essential that the segregation of waste is carried out to provide this material in usable form with maximum cost benefit to us.
- Part of our strategy is to redesign the product itself to reduce energy consumption in the manufacturing and use stages in its life. We also want to look at product weight as there may be some transport savings to be made here (and as a consequence energy savings too). The task is to devise some implementable options to do this, taking into account several key areas such as our customer needs, production schedules and product recycling/component reuse.

3 Motivating

Whereas instructing might be perceived as fairly cut and dried, motivating, persuading or encouraging might be necessary if an initial communication did not produce the intended outcome. As with most professional communication, you are looking for specific changes in behaviour which in this case will have measurable improvements in business and environmental performance. Change will be more sustained if staff are engaged and recognize the merits of the change for the business, the environment and themselves. Motivational communication might be:

- I know we've had some problems with transportation of our goods to the customer and how we package and store them. I'd welcome your thoughts and some specific ideas on how we can improve the situation.
- This year's data clearly show that the new working methods we have adopted have significantly reduced our energy and waste costs. Energy consumption is down by 8 per cent overall and we've reduced our CO_2 emissions by over xx and waste costs are down by nearly 13 per cent. As you will see from the paper I've circulated, this translates to environmental improvement too. I would like to say well done to you all for your considerable efforts over the past months.

4 Seeking

Here, instead of telling staff what is going on and how it will affect them, you are encouraging them to tell you. Asking the right questions, setting challenging tasks, or simply keeping

DOING

quiet at the right time in a meeting so others have space to bring forward their new and often innovative ideas will all help to achieve the desired result. For example:

- I think we could shave a significant percentage off our distribution costs by a combination of reducing packaging and perhaps increasing lorry payloads. I'd like you to look at this and report back at next month's meeting with some objective data on what we might be able to achieve.
- How much do we actually spend each year on waste contractors and what for? With this underpinning knowledge of the situation are there any other options we could look at?

Table 9.1 illustrates some specific uses of the four communication categories to draft your own 'communication needs' plan of action.

Table 9.1 Planning internal communication

Issue	Required outcome	Informing	Instructing	Motivating	Seeking
Waste	Improved segregation of waste on the shop floor		✓	✓	
Transport	Improvements to vehicle routings and so more efficient transport strategy	✓			✓
Energy	Reduced lighting costs by simply switching off unwanted lights	✓		✓	
The mix of categories can be extended for other issues and both internal and external audiences					

How do I implement a communication programme?

Having decided on the desired outcomes to match your priorities you now need to consider implementation. This can be considered in three categories:

1 Mix

To be successful in your strategy you will need to combine two or more categories of communication. For example, a production process might already use toxic or hazardous materials requiring compliance with legislation, so formal instructions will be essential. One EMS target might be to reduce the use of toxic substances and the process/product designers will need 'motivating' and 'seeking' as part of the mix. A less direct, but perhaps equally important, need might be to reduce energy costs in the production process. Here the mix might be a more balanced combination of all four categories. Getting the mix right will depend on the targets that have been set either by yourself to implement the EMS or imposed on you by external demands such as those of customers or regulators.

2 Media

Make sure you choose the right medium for your communications. A letter signed by the Chief Executive to every member of staff may be the best way of communicating that your new environmental policy is a significant document for the organization. Using in-house media such as notice-boards, newsletters and websites is a good way to raise awareness of your communication campaign.

3 Methods

A series of presentations, appended to existing staff meeting and briefing systems, will probably be the best way of beginning the communication process with functional groups of staff. It might make sense to prepare a short and punchy version of the report for all staff. Adding specific examples to match each audience group will help their identification with the overall strategy. You might also need to hold meetings with key staff who are directly involved in implementation of EMS (see Chapter 8).

Together, the 'Mix', 'Media' and 'Methods' define your communications plan for EMS. This needs to integrate with any overarching communications strategy for the organization as a whole.

How do I best deal with external communication issues?

The final part of your communications strategy is to consider external communications. You need to decide whether you are going to communicate externally about your significant environmental aspects (see Chapter 4) and if so how you will do this. While many stakeholders might have no interest whatsoever in, for example, your new production line, they may well have an interest in the environmental record of the business. This is especially true if customers themselves have established an environmental position. Your environmental impacts as a supplier will be a management issue for them.

The purpose of good external communication is to establish and maintain confidence and understanding among stakeholders. Many of the points already made about internal communications apply equally to external communications. External communication has an important role to play in your overall strategy, but may be subordinate to activity within the organization itself. You will know by now what the external demands are from your stakeholder analysis and you will know what your significant environmental impacts

DOING

are and the aspects of the business that contribute to them. If EMAS is your ultimate goal then there are some key requirements on external communication which you must meet. ISO 14001 also requires procedures to deal with external as well as internal communications. In addition, there is a further emphasis on communicating procedures and policies to suppliers and subcontractors. The 'Quick check' at the start of this chapter sets out these requirements. How you wrap this up into an external communication strategy will be a business decision only you and your colleagues can make. However, there are some general principles which will help formulate an external communications strategy. Once again, ask yourself some questions and consider the answers to them:

Why do we want to tell our stakeholders what we are doing?

Chapters 2 and 5 identified the range and interests of stakeholder groups. Many of these organizations and individuals are important players who can have positive and negative influences on your activities. A good communication strategy based on this work might well bring new business, improved insurance rates, and a profile which means prospective employees will recognize your organization as a leader by innovation. We can't think of better reasons for you to begin and maintain an ongoing dialogue with all your stakeholders about environmental issues. Better community relations will be a further benefit of a good communications strategy. An EMS denotes fundamental change in the way you do business. Continual change is founded by developing learning and understanding between you and your suppliers and customers so that your and their needs continue to be met in a realistic way.

What do they want to know?

Stakeholder needs, like stakeholders themselves, will vary. Some may want to know very little about your business, so long as it is legal and showing a profit. As part of your proactive communication strategy it is incumbent upon you to investigate, understand and respond to the diverse needs of stakeholder groups.

What do we want them to know?

Above all else tell stakeholders about your positive activities past and present. People don't just want to hear about your good intentions, no matter how grand they may seem on paper. The holistic approach of EMS and its realization through effective management represents a radical departure from more traditional reductionist approaches. There are powerful messages which illustrate that the benefits of the approach accrue to the business and all stakeholders. These messages represent opportunities to reaffirm the organization's stance and to lead by innovation. If you decided it makes good business sense to communicate externally about your significant environmental aspects, record this decision and consider just how you will do this.

How can we communicate with them?

Think carefully about your options in terms of the ways of communicating. You can control absolutely what you write in a press release or annual report, but once it has been transmitted to the reporter, his or her own private bias or the publication's own agenda may well camouflage or reinforce your words. Don't forget your own staff. If you are serious about implementing an internal communication and training strategy then, whether you like it or not, each member of staff effectively becomes a spokesperson for your organization. People believe what their peers, colleagues and friends tell them. If you were a fly on the wall outside the workplace what would you like your staff to say about your organization's environmental position? Various routes exist to establish effective external communication. All require you to take the initiative by building direct relationships with your stakeholders and constituents. Dialogue is two way communication after all. You might start, for example, by setting up supply chain groups or inviting customers to sit in on your in-house environmental training courses.

Table 9.2 Considerations for the communication programme

1 Consider your own motives and integrity		
Question	Action?	Thoughts
Do I have respect and understanding from my colleagues?		
Am I definitely concerned about environmental management or am I doing it simply to boost my own image?		
Do I practise what I propose to teach?		

2 Understand the person(s) you are going to be working with		
Question	Action?	Thoughts
Against what criteria would you justify EMS work in terms which my colleagues in other organizational functions will understand?		
Ideas include: • in terms of morale • in terms of sales, external perceptions and publicity? • in terms of technical aspects of the organization? • in terms of cost savings or cost neutrality? • in safety terms? • in terms of legislation? • in terms of individual aspirations? • in terms which will inspire staff to act?		
Can I justify and demonstrate the benefits of EMS in ways which make sense to the business?		

DOING

Table 9.2 *(continued)*

3 Underpin decisions rigorously		
Question	**Action?**	**Thoughts**
Am I planning EMS implementation well and in time?		
Am I matching my strategy with existing in-house mechanisms?		
Is my communication and persuasion of opponents ongoing?		
Do I emphasize and promote the positive at every opportunity?		
Do I make sure collective decisions are followed through?		
Do I always acknowledge positive individual and team actions?		
Do I back up my colleagues if they come under fire?		
Do I always say thank you?		
4 Set priorities		
Question	**Action?**	**Thoughts**
Do I have a short, medium and longer term strategy for the sequencing of EMS implementation?		
Is my strategy realistic given my understanding of the organization and the people in it?		
Are the decisions I want re EMS set out in a realistic progression?		

5 What we have to do: EMS requirements		
Action	**Action?**	**Thoughts**
Set out a schedule of 'must do' requirements aimed at implementing EMS in the organization		
Set out to gain unanimous support from management for these 'must do' actions by illustrating that: • a systematic EMS strategy will benefit the business • EMS, like quality, requires total commitment • EMS is a journey not a destination		
Ensure that all senior management are aware that the environment must always be on the agenda at the highest level in the organization		
Reinforce the unanimous decision for continuous improvement through measuring and monitoring		

Table 9.2 *(continued)*

6 What will benefit the organization?		
Action	**Action?**	**Thoughts**
Set out an inventory of EMS measures that will bring tangible benefit to the organization		
Present the measures in ways which reinforce the positive aspects		
Ensure these measures are implemented		
Work to illustrate EMS measures are essential and will bring direct and indirect cost benefits		
Report on success stories once you have hard data to tell		

How do I motivate people?

Changes can only be effectively introduced with the accord of the organization's management team. Any members of the team who are not yet convinced of the need to implement an EMS will need to be convinced lest they undermine its realization. In most organizations this commitment to change has to be sanctioned by senior management if it is to survive. There are six areas to reflect on before starting a communication programme; these are detailed in Table 9.2.

As Georg Winter notes in his seminal publication, *Business and the Environment*: 'tell the truth, be clear and keep your actions consistent with words. . . It is better to put money into good work rather than into good reports on mediocre work'. Effective communication is a central component of the strategy to stay ahead of the competition. In the next chapter we will look at an important manifestation of communication: training and awareness.

 Narratives

Smallco

The experiences in implementing EMS management programmes to deliver the objectives and targets set have highlighted a real need to cut across formal communication channels. They simply don't work that well. Staff tend to model what is explained and demonstrated to them systematically rather than through their own interpretation of two-dimensional paper-based communication. In the case of the EMP staff focused on what they felt was best rather than what they had been requested to do in an internal memo or work instruction.

For the Ops Manager, the learning from this experience and its solution – establishing regular verbal communication with staff about the EMS – set the scene for effective internal communications in Smallco. The outcome, and what worked, was face to face meetings with staff on a regular basis. Staff responded best when they were invited to take part by a person not a piece of paper. The leading edge for internal communication was personal communication underpinned by any documentation that was deemed necessary.

DOING

External communications, begun with the release of the environmental policy, have gone very well. The developing working relationship with the major customer has helped by providing both resources and a sounding board and helped Smallco fast track to an effective communications strategy.

BIG Inc

Because of the strategic nature of much of BIG Inc's objectives, it is not clear who needs to have what information at what stage of the process. As a result, the minutes of meetings and other internal communication documents are virtually broadcast to everyone using the company intranet. Unfortunately, this has the opposite effect of what is intended. Because everybody is seeing everything, the system is becoming overloaded and nobody has the time to discover what is relevant to them.

This prompts the project manager to ask the Corporate Communication Adviser to initiate a review of the internal communications function. The review divides all communication into four categories:

1   Direct (face to face and interactive);
2   Indirect (impersonal using the intranet, publications or even group voice-mail);
3   Open (broadcast to all);
4   Closed (specifically targeted).

Using the results of the review, managers are advised that at the end of each meeting on environmental matters, decisions are to be taken as to the most appropriate route for the information. Gradually this helps the company to move from over-reliance on open indirect communication towards closed direct techniques and in such a way that use of the intranet itself can give feedback on the progress being made by measuring types of usage.

Trade secrets

- Practise what you preach, and do all you can to demonstrate good practice not just talk about it.
- Set high standards, ideally you should all speak the same language.
- Make sure your staff have good things to say about environmental practice, ideally something they are directly involved in.
- Become a listening manager and hear what your staff say.
- Go on 'walkabout' as much as you can and observe what is really going on.
- Always reward good practice with professional praise.
- With external communications, make sure you know who you are speaking to and what they want before you open your mouth.
- Tell the truth. It is easier to climb out of a small hole than to dig and then try and escape from a large one.

⍰ Things to think about

- Expect *ISO 14063 Environmental management – Environmental communication – Guidelines and examples*, to become a watchword in environmental communications. It is a very useful point of reference for further guidance as you develop your communications strategy.

- Sustainability and governance issues all look to the use of stakeholder dialogue as an ongoing and effective technique for keeping an organization 'on track'. Anyone with environmental communications experience will be worth their weight in gold to organizations starting down this route.

DOING

10

Competence, Training and Awareness: Do We Know How To Do This?

In previous chapters we have considered the basis and setting for an EMS. This chapter will provide the framework for a dedicated training, communication and awareness programme to support the EMS and ensure personnel dealing with environmental issues are competent.

✓ ISO/EMAS quick check

Area of EMS	ISO 14001	EMAS
Competence, training and awareness	4.4.2	Annex I-A.4.2

▢ Chapter executive summary

Move to the next chapter when you can answer all of the following questions in ways which make sense to you and your organization:

> *How can I establish competence and training needs?*
> *How do I link competence and training needs with identified environmental aspects and EMS needs?*
> *What is the best way to implement a training programme?*

⤲ You are here

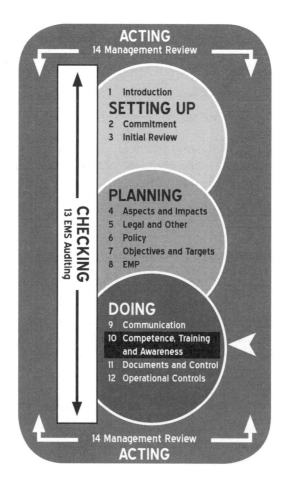

⚒ Toolkit requirements

To work through this chapter you will need the following knowledge and skills:

Knowledge

- organizational structure and existing training practices;
- what EMS means for your organization;
- past and planned training;
- stakeholder requirements;
- environmental aspects, related competences and organizational priorities.

Skills

- matching skills (what needs to be done with how to do it);
- interpretation and planning skills;
- interpersonal skills;
- presentation skills.

⚲ Context

Before developing the use of training techniques to reinforce the EMS process by developing competence it is important to consider the nature of training itself. Two definitions provide guidance:

> Training: A planned process to modify attitude, knowledge or skill behaviour through a learning experience to achieve effective performance in any activity or range of activities. Its purpose, in the work situation, is to develop the abilities of the individual and to satisfy the current and future manpower needs of the organization (UK Manpower Services Commission, Glossary of Terms).

Or to put it more simply, training is:

> Any activity designed to improve another individual's performance in a specified area (Malcolm Peel, UK Institute of Management).

The important feature of both definitions is the paramount objective of training to make positive changes to individual performance by developing their competencies. Effective organizational change, vital for the realization of an EMS, will flourish from such intervention. Training for EMS presents new challenges to the training process. It will require the provision of training both for initial orientation and as an ongoing process to manage identified environmental aspects.

Consider the following statements. How do they relate to your organization?

Training and communication are essential to:

1   Equip all staff, especially those whose role(s) are related to identified significant environmental aspects, to perform their jobs with competence. *In the context of EMS this might mean there are changes to work procedures to embrace objectives and targets set in Chapter 7. So, for example, staff need to know more than simply the fact that monitoring of emission is needed, they actually have to be competent to carry out the monitoring tasks.*

2   Provide and maintain the knowledge to enable staff to understand why their actions count. *There are many external motivators for environmental management – covered in earlier chapters – which impact on the business and therefore the employees too.*

3   Promote employee identification within the organization. *The realization of a purposeful EMS will require innovative cross-functional working which can only come from strong identification.*

4   Increase employee commitment and motivation. *Organizational and cultural inertia favours 'business as usual'. Environmental management may sometimes require 'business as unusual'.*

5   Avoid failures. *A fundamental element of an effective EMS is to ensure the organization manages its environmental aspects which might result in significant environmental impacts, and as a result preventing pollution and striving for continual improvement.*

Training, and the communication that underpins it, is about skills transfer, motivation, changing attitudes, developing understanding and the bottom line – action which makes a positive contribution to the organization and the environment. Alongside the actual mechanics of, say, monitoring to ensure regulatory compliance, employees will achieve competence if they fully understand the decisions that affect them, how and why these decisions arose in the first place and how their professional contribution will actually make a difference. Developing a sound understanding of staff role is a key component of effective EMS implementation and management. In essence, a training, communication and awareness programme should aim to identify and develop environmental management literacy and practical competence throughout the workforce.

   Involving all staff in implementing good practice and management of their contribution is paramount. Sustained change in individual behaviour and professional action will only be achieved through effective training and communication; both form a continuous process throughout the life of the EMS.

 Tasks

How can I establish and manage my organization's competence and environmental training needs?
It is not enough simply to acknowledge the benefits of training. Successful training must always change how people behave. Effective training is the catalyst for the process of realizing strategies adopted in earlier sections of this book. For effective EMS implementation and ongoing management, three areas need to be considered as shown in Table 10.1.

Strategic needs

The most important first step in an analysis of environmental competence and associated training needs is to create and develop the 'big picture' for organizational needs. This analysis of strategic needs is a much more effective starting point than a fragmented analysis at the level of the individual. The starting points for strategic analysis of needs should be the business plan for the organization, the EMS implementation plan and particularly EMS objectives and targets which reflect environmental aspects and associated impacts of workplace activities. All of these are dynamic and include:

1   The organizational responses to motivations for environmental management:
   (a) customer expectations;
   (b) environmental stewardship;
   (c) sustainable development and corporate social responsibility (CSR);
   (d) environmental regulations;

DOING

(e)   risk management;
(f)   eco-labelling;
(g)   voluntary standards;
(h)   other stakeholder drivers;
(i)   competitive advantage.

Table 10.1 Training areas for EMS implementation

Training area	What is required?	Where is the information?
Strategic	An examination of the overall picture for the organization	The organization's business plan, company environmental policy and commitment and the EMS strategy itself
	Long- and short-term organizational needs	Discussions with senior management Sight of the business plan if available Established EMS objectives and targets, see Chapter 7
Functional	An examination and identification of specific functional training needs, for example, the marketing department may need new knowledge and skills to promote your environmental position	Existing knowledge of product and process and new goals and targets Supplementary interviews with managers, staff and customers in identified functional areas
Individual	Identification of individual training needs to deliver the EMS strategy	New objectives and targets including individual ones and IER work
	Awareness for all	Will also require a more detailed analysis which may use existing personnel and management systems

2   Organizational aspects of EMS. These elements form the internal framework for EMS and include:

       (a)   overall business targets;
       (b)   environmental objectives and targets;
       (c)   what might help your business;
       (d)   environmental policy;
       (e)   management strategies;
       (f)   your initial environmental or baseline review.

Although human resource implications might not be explicit in these plans, they will exist. For example, the introduction of a new product may need staff with new skills to manage and operate the production process and ensure performance objectives are achieved. The new product may realize economic and environmental benefits by minimizing resource depletion and waste generation throughout the production process. For the organization the choices may be to recruit staff with new skills and competences, retrain existing staff or a combination of both. Those involved need to be trained to ensure competence and the best return on investment in the new plant and to insure against any negative aspects of not embedding required competences.

A rigorous examination of the business plan, the EMS implementation or its ongoing management plan coupled with the knowledge gained in the IER, and setting objectives and targets in earlier sections will signpost needs. There are several questions to keep in mind as you address these issues. The list below will provide a useful aide-mémoire:

What is already 'on paper'?
What are the competence and training implications of the organization's existing business and other plans?
What is being done well?
Where might the organization be potentially exposed now and in the future?

What is the current level of understanding?
Are staff aware of environmental implications of their work?
Are new skills and competences required?
How will they be obtained?

What is already underway?
Is the organization already managing some aspects of its activities which have an environmental component (eg waste and energy)?

What changes will be needed?
Will an EMS require future functional changes which will require training interventions to ensure competence?

What part will training play?
Does the EMS 'plan' imply a culture shift – for example, towards, proactive working (empowerment) – which may require training interventions?

Functional needs

The second step in analysing competence and training needs is to identify functional issues. You will now need to extend your knowledge with supplementary interviews with managers, staff and customers in identified functional areas. It is essential to take a representative sample of staff in each functional area of the business to ensure the findings will give an unbiased picture of what the needs are. For example, staff will not all have had the same experiences before joining the organization. Some may be experienced operators while others may be relatively new to the organization. Judging needs based on a sample of one and not taking account of diverse needs is a sure way to set up a programme which will fail.

An in-depth training needs analysis (TNA) is needed to identify functional training needs. Table 10.2 begins this process.

DOING

Table 10.2 Identifying functional training needs

Question	Comment	Your thoughts . . .
What do you understand by the term TNA?	The term training needs analysis encompasses a variety of techniques ranging from the informal and often subjective to rigorous analytical interview techniques designed to detect detailed information	
How might you set about beginning a TNA? Some questions to guide this process are:	Which functions and departments have greatest need? As derived from your work on IER and aspects and impacts	
	What are these needs?	
	How would staff actually meet performance measures explicit in your objectives and targets for the EMS?	
	What training is already provided? Is it sufficient to meet identified EMS needs?	
	Are staffing levels right for the job? Is staff turnover an issue? Are there training implications?	
	Are any changes proposed in operational practices? What are they and what will they mean for the operators and the realization of environmental objectives and targets?	
	Where are skills and knowledge gaps which will need to be bridged to ensure environmental competence for implementation of the EMS?	
	How might training effectiveness be measured?	

Individual needs

While the first two steps were, in the main, concerned with the core and planned changes in business functions, this step engages the activities of the individual, their awareness and aspirations. Staff whose role has the potential to cause a significant environmental impact need to be aware of the whole EMS, the issues that the organization is addressing and how their actions can influence and improve the performance of the whole business. Within any organization, a number of procedures may already exist, which may directly or indirectly reveal competence and training needs. What evidence would you expect to find for each of the issues given in Table 10.3?

The environmental management representative: A special case

A range of skills are required by the manager with responsibility for the EMS itself. These range from the hard competences as set out, for example, in ISO 19011: 2002 relating to

Table 10.3 EMS training needs

Issue	Any identified EMS training needs?
Appraisal/Performance review	
Test and examinations, for example, arising out of competence-based training	
Self tests	
Assessment of prior learning	
Mentoring	
Individual career planning	
Career counselling	

auditing skills (see Chapter 13). This body of knowledge and skills (concerning relevant science, technology, industry and process understanding, environmental law and regulations, EMS principles and the audit process) might not all be utilized in a practical way but will be required for the EMR to perform their professional role effectively. Other relevant knowledge and skills relate to project management, managing change (see also Appendix III) and the interpersonal skills needed to carry out the diverse requirements that analysis of communication and training needs will undoubtedly reveal. These include such areas as basic interpersonal communication; the ability to delegate; the ability to motivate; team building; interview skills; presentation skills (both oral and written); report writing; negotiating skills; time management skills; resource management skills; and a good understanding of financial management. It is easy to overlook the training needs of the environmental manager themselves, especially if they are actively involved in the rest of the work to install and manage an EMS. Acknowledging that the environmental manager has tangible training needs is an important step to take. Their pivotal role in making the system work will mean that they too have skills to develop.

In this first section of the chapter we have discussed needs analysis in some detail. The next phase of activity is to develop focused training interventions to ensure those that need new or refresher skills and competences to implement the EMS actually acquire them.

Training objectives

Training is not simply about providing information. It is about effective management of change and involves a combination of skills development, knowledge transfer, development and support of understanding among staff and, most importantly, sustained attitude change. All these ingredients will be required if the training programme is to meet the identified and ongoing needs of the EMS. Training objectives for EMS:

Area	*Objectives*
Competence development	To bridge identified skills gaps to ensure new skills and competences are acquired. For example, skills identified earlier to develop the individual's ability to put into practice EMS requirements in their own work situation.
Knowledge transfer	To ensure knowledge required for effective management of EMS is transferred to relevant staff. For example, that required to manage aspects to reduce significant environmental impacts.
Development and support of understanding	Through training activities, to develop a working environment and cross-functional relationships which will amplify the beneficial aspects of EMS.
Sustained attitude change	To ensure that training interventions challenge existing perceptions by raising fundamental questions about environmental management.
Awareness for all	To help staff understand and deal with change represented by the EMS in many forms.

Before embarking on a training programme – for an individual, functional unit or a whole department – you will need to consider the position of your learners both pre-training and where you want them to be by the conclusion of a training intervention. Think carefully about any barriers to change. What is realistically achievable in the timeframe you have set or which has been imposed, and what resources are available? Training objectives will need to address these broad aims and reflect a number of EMS issues. Use the space in Table 10.4 below to add brief notes on your own situation.

Table 10.4 Training objectives

Generic need	Specific need
The knowledge that will be needed by all staff to underpin an EMS training programme	
The skills staff will need to effect change. Some skills will already reside in the business while others will need to be introduced through training	
The attitudes that it will be necessary to develop among staff if they are to be active participants in the implementation and development of an EMS	

✓ Active participation is an essential part of the learning process.

✓ Good learning takes place when adrenalin flows.

✗ Passive 'training' activities, for example, watching a video then being asked to talk about the relevant issues is a poor substitute for a dynamic training process.

✓ Effective training engages participants and their skills in an active way.

✓ Training outcomes must ensure organizational objectives are met *and* individual satisfaction and success are achieved.

How do I link training needs with EMS needs?

Establishing training needs in the three areas – strategic, functional and individual – will begin the process of devising an organizational training programme. One of the outcomes of this work will be a mapping exercise which begins to highlight training needs and EMS. For any of the TNA findings it is possible to run through steps 1–4 of the training cycle below and determine which EMS priorities require a training intervention to improve performance.

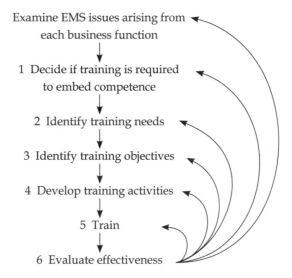

The feedback loops are an important component of the process and a component of measuring training effectiveness. They will be discussed later.

Consider just one key EMS issue for your business and select one aspect which requires a training intervention to ensure competence among key staff. Think about how you might address each step in the training cycle using Table 10.5. Now consider any feedback loops appropriate to the example you have selected. Are some feedback mechanisms more or less important in your situation? Why do you think this is so and what might the management implication be?

Table 10.5 The training cycle

Examine an EMS issue	Training
1 Decide if training is required	Yes/No/Why?
2 Identify training needs	
3 Identify training objectives	
4 Develop training activities	
5 Train	
6 Evaluate effectiveness	

What is the best way to implement and manage a training programme?

Methods

However well training needs have been established, the training interventions will only be effective if suitable training methods are selected. For example, the operation of a new production process to reduce raw materials consumption must be demonstrated by practical activity, whereas training to encourage office energy management might best be carried out by a combination of lecture and workshop sessions which includes practical activities to quantify the difference that can be made. Earlier sections of this chapter examined priorities and training needs. The outcomes were categorized as requiring changes in knowledge, competences or attitudes. In the most effective training interventions all three outcomes will happen simultaneously with emphasis on each component balanced to the actual training goals. The training methods you actually use will depend on these desired outcomes.

Training methods themselves can be categorized into in two main areas: individual training and group training. These are summarized in Table 10.6.

Routes

Your TNA will have revealed the training need. By linking it to your work on environmental aspects and impacts, the objectives and targets for your EMS will begin to highlight areas for action. Many opportunities will exist to make training interventions. These are summarized in Table 10.7.

Evaluation and effectiveness

Learning from experience and applying this learning is a vital part of effective management. To be effective, this needs to be done through regular evaluation of performance. To ensure your systems and the programmes to manage them connect seamlessly, two competences are needed: the ability to measure and monitor performance. These competences apply to all aspects of the management of the EMS, including the HR issues of training and communication.

Table 10.6 Comparison of training interventions

Individual training methods	Cost	Notes
Structured reading	Low cost	Can be a useful technique alongside other activities but may not be appropriate for all trainees
Open learning	Medium cost	Requires strong self motivation from the learners. The best programmes combine knowledge-based activities with work-based practical activity
IT-based learning	Medium/high cost	One of the most quickly expanding training areas, often using CD-ROM. The best materials can fully engage the trainee at their own pace and provide feedback for the trainer
Coaching and mentoring	Medium cost	These methods provide support and first hand interaction with more experienced staff and can reinforce other training interventions
On-the-job training	Medium	Can be an effective technique for activities which combine knowledge, skills and techniques. Very dependent on the abilities of the instructor
Group training methods	Cost	Notes
Lectures	Low	One of the least effective training methods with 20% retention. Trainees' attention span often much less than length of lecture. Best suited to general overview and introduction of new ideas
Group discussions	Low	Trainees much more actively involved than lectures. Syndicate groups can ensure full participation of all and reporting back an excellent way to test understanding of new knowledge attitudes
Role playing	Medium requires higher trainer/ trainee ratios	Because role playing requires trainees to simulate real work behaviours it is a powerful technique to develop interpersonal skills. It can help bridge the gap between theory and practice. May exclude some trainees because of their reluctance to participate
Audio-visual	Medium/high	A well-produced audio-visual presentation or video will maintain trainees' attention. However, a commercial production may not reflect the organization's precise requirements and may be perceived as irrelevant. Too often, a video can be used as a light break in a training session rather than as an integral part

DOING

Table 10.6 Comparison of training interventions *(continued)*

Group training methods	Cost	Notes
Case studies	Medium	This training intervention can stand alone or be integrated with role play. It is an effective method to engage trainees in activity which specifically relates to the organizational needs
Games: board and computer-based	Medium	Can be an effective training tool if well integrated with the overall training strategy
Outdoor training	High	Primarily to develop team cohesion and skills. Ideally they should combine several of the training techniques noted here as well as outdoor programmes

For a detailed analysis of training principles and practice see M. Reid and H. Barrington. (1994) Training interventions, IPD.

Table 10.7 Opportunities for training interventions

Training route	Notes	When can I use it and what for?
Pre-joining	A copy of your environmental policy might well form part of the invitation for interview pack	
Induction	An early opportunity to set out the organization's environmental game plan and expectations of staff in its implementation	
Team	Identifiable 'teams' will exist in the organization and working with them in their work setting is a productive training route	
Department or section	Working with a department or section will allow a clear focus on their specific needs which may differ from other sections of the business	
The whole business	There are many opportunities to reinforce systematically the EMS messages throughout the organization. These could include internal vehicles such as staff newsletters and notice-boards (report successes) or external such as the media. It is perhaps here that training and communication blend together	

Ask yourself four questions about any training intervention:

1   Does my training programme achieve the objectives set?
2   Does my training improve organizational performance?
3   Was my training worthwhile?
4   What will I do next time to ensure my staff are competent?

This section looks at how to find answers to these principal questions.

Measurement

Measurement is an essential component of both maintaining and improving EMS performance and was dealt with in detail in Chapter 8. Effective implementation of EMS demands effective measurement of performance. Controlling data is a key task. Objective evidence in the form of hard data provides the baseline against which progress can be measured. The setting of objectives and targets discussed in Chapter 7 provides the framework for performance objectives. These can be qualitatively and quantitatively measured and mapped against any training interventions. Two examples are shown below:

Objective	*Performance measurement*
To train BIG Inc operatives in new handling techniques for volatile organic compounds (VOCs)	Measurable reduction in use and purchase of VOCs over a defined time period
To train Smallco's shop floor staff in new waste segregation procedures	Reduced waste to landfill; reduced costs for skip hire; better prices for scrap metals

Monitoring

Like measuring, monitoring signals a commitment to the management of the EMS and the development of a positive organizational culture. Actual areas to be monitored are as above for measurement. To be effective, monitoring needs to reflect the realities of managing an organization. Two types of monitoring – active and reactive – will ensure the day-to-day realities of organizational operations are addressed.

1   *Active monitoring* = preventive action. Active monitoring is the essential feedback on performance against EMS objectives. It involves checking compliance with performance standards – those you have established for yourself or those imposed by regulations – and the achievement of specific objectives. Its key purpose is to demonstrate success.

2   *Reactive monitoring* = corrective action. Reactive monitoring manages deficient EMS performance, for example, direct non-compliance with established EMS targets or breaches of legislation.

These competences – to measure effectively and monitor performance – are promoted by training and communication, the goals of which are to create an organizational culture

DOING

emphasizing a diligent approach by all staff to the implementation of an EMS. They must be presented so that they can be interpreted by management to allow appropriate actions when these are required.

Effective training

Training can slide into becoming an act of faith which it is hoped will achieve some benefit. These subjective judgements are an inadequate gauge of quality. The need for effective EMS management requires a much more rigorous set of techniques to evaluate training programmes, and the provision of hard data on which to base management decisions. Evaluation is an assessment of the real, objective value of a training programme in both organizational and environmental terms. Think about the following statements:

- We knew we needed to do something about training when we breached legislative requirements twice in three months.
- Our training programme is based on ensuring that at all staff receive specific, focused and continuing training corresponding to their role(s) in the management and implementation of our EMS. I know they are competent to do their jobs.

Why do you think that the second approach is more likely to help the organization?

The mapping of training needs against the organization's EMS strategy will provide a programme with clear, outcome-based training objectives. With these clear goals it is possible and desirable to measure the effectiveness of training. An evaluation model can be constructed using five components:

1   *The response of the trainees to the training.* Asking trainees questions about the environmental training they have undertaken provides valuable feedback both long and short term. The responses of trainees are important because they indicate an attitude to learning and motivation. Your efforts to sustain a positive attitude through training programmes will enhance the learning. Ranking the answers on a scale of 1, negative to 5, positive, will provide comparative data and can suggest fine tuning of a training programme.

2   *The learning achieved.* Gain analysis – testing knowledge, skills and attitudes before and after the training – will provide a measure of the effect of the training. Having clear objectives before starting training is vital; hence the importance of TNA and mapping this against requirements emerging from the organization's EMS strategy. The key questions to ask yourself are: Where are we now? Where do we want to get to? How will we know when we have arrived?

3   *Changes in trainees' behaviour.* Simply because trainees enjoy the course, does not mean that the new knowledge and skills will automatically transfer to the workplace. Many barriers exist to this transformation

ranging from opposition of colleagues to change, through to the trainees themselves being unsure of how to apply the learning. To apply objectivity here requires planned observation of the trainees' real job performance and where necessary to support the transfer of skills to workplace practice.

4   *Organizational effects.* While training interventions are assessed on improving the performance of individuals, it is important to relate training to both the functional and the strategic changes identified to achieve effective EMS management in the organization. You will need to evaluate the effectiveness of training at the organizational level. This is no easy task. The specific effects that training has made on the implementation of EMS at the organizational level may be concealed by other factors such as global changes in work practices or directives from senior management. Training is an important component in the change mix and every effort should be made to identify and quantify its contribution.

5   *Real value.* This last component of the evaluation model transcends basic questions such as 'how much did the training cost?' or 'how well did the programme work?'. It has more to do with philosophical position and outcomes in terms of value and attitudes of the organization and its place in the wider society. The economic and environmental benefits of an effective EMS are shared by multiple stakeholders as well as by the organization itself.

The five stage evaluation model provides an overview of training evaluation and the rationale for doing so. The actual measurement of training can be carried out by thinking about short-term effectiveness, long-term effectiveness and cost effectiveness.

Short-term effectiveness

Most training interventions seek to produce changes in behaviour. This can be measured before and after training. For example, a waste management programme may mean consideration of the entire supply chain and production processes within it. The chain can be broken down into its component parts. Each is analysed to reveal the information and procedures staff need to understand and how this information will be conveyed to them. Performance and competence can be measured before and after any training interventions. Measurement can be made more accurate by introducing precise objectives. For example, compare:

We wish all our staff to become effective in environmental management techniques.

with:

For an employee to be considered competent in the use of environmental management techniques in their job means:

1 Full cognizance of the organization's EMS strategy, policy and the objectives and targets relevant to their role.

2 Identification of EMS tools and techniques relevant to the task and job function.

3 An understanding of environmental aspects and impacts associated with their job.

4 Demonstration of excellent integrated cross-functional working methods.

5 Establishment of appropriate performance review techniques.

Clearly, the second set of objectives provides for more rigorous measurement and monitoring.

A number of generic questions may help in designing measures for training effectiveness and short-term gains:

- What specific objectives were set to improve environmental performance?
- Have they actually been met?
- Is there an internal or external test or examination of these objectives?
- What measurements of performance can be made?
- Who needs to be involved or considered in the evaluation:
  - the trainees?
  - professional peer group?
  - the managers directly involved?
  - any other internal stakeholders?
  - external stakeholders?

The performance of individuals can be assessed at four levels as shown in Table 10.8.

Long-term effectiveness

If not practised and supported, new skills and behaviours are abandoned and old, established ways of behaviour quickly become the norm again. Ineffective training, which only partly develops skills and behaviours, will not transfer the need and ownership for them to the trainees and fails to integrate the learning with the learners. The outcome is trainees who lack confidence. These are all good reasons for re-evaluating any training to ensure its long-term effectiveness as the EMS evolves over time.

The actual definition of long term will depend on the situation, but it is likely to be between 3 and 12 months following any training intervention. The methods for long-term evaluation are just the same as those discussed for short-term evaluation above.

Cost effectiveness

Training for its own sake, because it is a 'good' thing to be doing is no basis on which to commit resources. Such an approach reinforces a negative attitude to training and is somewhat hit and miss. Effective EMS implementation and management demands a systematic and holistic approach. Managing training means:

Table 10.8 Individual performance assessment

Level	Assessment options
Skills eg ability to implement relevant parts of EMS	observation simulation exercises pre-defined practical testing
Knowledge eg understanding of policy or implementation strategy	training assignments oral questioning multi-choice questions objective questions written examination
Attitudes eg relationship with other units	observations questioning this and other units questioning external stakeholders to determine if perceptions have improved
Ability to apply skills/knowledge eg ability to implement EMS in the workplace	analysis projects/assignments reports recommendations presentations

DOING

- Demonstrating in the terms of the trainees, that training, and associated improvement in competence, is a key component for the implementation of an EMS. The 'Quick check' at the start of this chapter identifies the standard requirements for training.
- Developing an understanding of the investment, including staff time, lost production as trainees are off the job, and the range of monetary costs of training such as consultancy support and material resources.
- Knowing what the return is on this investment; a cost benefit analysis (CBA) of training using measurement and monitoring methods discussed earlier.

Alongside this work, the actual cost of training needs to be established. This is a straight-forward budgeting operation, with budget lines for each element of the training activity. A training budget should include lines for:

- capital: the training centre and associated facilities;
- staffing: trainer(s), trainees, staff cover and the supporting administration; and
- equipment: machinery, materials.

For each you will need to identify whether the costs are:

- fixed: independent of the level of training activity;
- variable: dependent on the number of trainees; or
- marginal: the additional costs of one more unit of training.

The cost effectiveness of training can then be derived from these two sets of data; on one side how much the training cost and on the other side what gain was generated as an outcome of the training process. This is vital management information.

For many organizations, the cost effectiveness of training means more than monetary gains. Training to implement EMS will help provide less easily quantifiable benefits such as competitive edge and improved staff motivation and retention. Organizations that don't train their staff and commit to lifelong learning risk stagnation. The systematic nature of EMS may well represent a radical departure from usual business practices. Organizations that train effectively, relate training to the core business activities and set carefully derived objectives for staff will, over time, change their own culture. Perhaps more than any single issue, culture change – whether at organizational or operating unit level – is paramount for a sustainable EMS.

 Narratives

Smallco

In the early months of their phased implementation of the EMS, on the job training was provided by the Ops Manager or individual managers. Although not particularly systematic, this has worked quite well so far. However it is clear to the Ops Manager that this can only be an interim measure. Ongoing implementation of the EMS will require training interventions above and beyond on-the-job briefings.

The Ops Manager decides it is time to carry out a step-by-step review of the training needs of the EMS itself. The established objectives and targets, and the EMP rooted in these, will require new training interventions. By doing this and then examining the actual expertise among staff, the Ops Manager working with the Quality Manager (who also has HR responsibility within Smallco) are able to identify the gaps between EMS needs, staff skills and internal training expertise. Training budget and staff time away from the job of production are both very limited, so the offer of help by the major customer to provide training where there is common interest in improvement, for example, waste minimization and packaging of product has been a great help. Never slow in coming forward, Smallco have quickly latched on to the offer. This systematic approach means resources and effort can be focused where there is need, and the EMS implementation kept on track more effectively than allowing a piecemeal approach to become established practice within Smallco.

BIG Inc

Being a young company, and having a management team that was involved in the first wave of environmental issues in the 1970s, awareness of environmental issues generally within the organization is high. Unfortunately, connecting these issues to what goes on in the organization is another matter.

Using a series of facilitated workshops – rather than simple presentations on the subject – means that all levels of company staff can feel as though they have made a contribution, tested their own understanding and begun to link their actions in and around the office with the EMS objectives and targets. At first, there seem to be strong moves towards staying with the obvious work-related areas, such as recycling office paper and using recycled ink cartridges in the computer printers.

Owing to the inclusion of many of the associates in the programme, especially those who are home based, the staff begin to explore the whole idea of all their actions impacting on the environment, both directly and indirectly. Staff morale and involvement with work seems better than ever after these workshops and a reduced version of the outputs is included in induction material. Graduate recruits, vital to the company's expansion plans and continuing high performance in their competitive market, report that the company's approach to EMS was certainly one of the elements that attracted them in the first place.

Trade secrets

- Training is often thought of late in the day or ignored completely. Those organizations that invest in training report measurable benefits.
- Focus carefully on what your organizational training needs are, not what a training provider is offering in a course prospectus. Inappropriate training has the potential to do more damage than not training at all.
- Integrate training with your whole EMS. It can be regarded as a day or two off work and a bit of light relief, so provide follow through activities to ensure the training gain moves into the workplace where it is actually needed.

Things to think about

- Sustainability and CSR initiatives are on the increase. By their very nature, they tend to be trans-disciplinary rather than restricted to one management discipline. Training and continuous professional development can either help or hinder such organizational development so be aware of where the broader perspectives prompted by environmental engagement might lead.

DOING

11

Documents and their Control: Where Have We Been?

Documentation should always be kept to the efficient minimum, though perhaps this is easier to say than to do. Used like maps, documents can help to keep you and your company on track. Used as records, they can help to show where you have been and how you got where you are. Used as an excuse for thinking properly about the provision of objective evidence that the system is being used, they quickly become wallpaper which is pretty to look at but merely papers over the cracks.

✓ ISO/EMAS quick check

Area of EMS	ISO 14001	EMAS
General requirements	4.1	Annex I-A.1
Environmental policy	4.2	Annex I-A.2
Objectives, targets and programme(s)	4.3.3	Annex I-A.3.3
Resources, roles, responsibility and authority	4.4.1	Annex I-A.4.1
Communication	4.4.3	Annex I-A.4.3
Documentation	4.4.4	Annex I-A.4.4
Control of documents	4.4.5	Annex I-A.4.5
Operational control	4.4.6	Annex I-A.4.6
Monitoring and measurement	4.5.1	Annex I-A.5.1
Evaluation of compliance	4.5.2	Annex I-A.5.2
Non-conformity, corrective action and preventive action	4.5.3	Annex I-A.5.3
Control of records	4.5.4	Annex I-A.5.4
Management review	4.6	Annex I-A.6

📄 Chapter executive summary

Move to the next chapter when you can answer all of the following questions in ways which make sense to you and your organization:

> *Will you know when you have written enough procedures?*
> *Do you know how to go about writing procedures?*
> *Do you know how to control documents?*
> *Do you know which elements of your EMS are affected by documentation and its control?*
> *What are the extra documentation requirements in EMAS?*

⮧ **You are here**

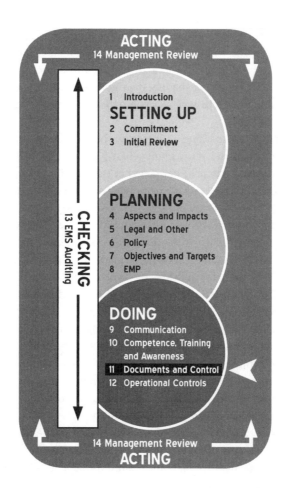

⚏ **Toolkit requirements**

To work through this chapter you will need the following knowledge and skills:

Knowledge

- range of operational controls relative to the significant environmental impacts;
- range of procedures needed;
- risk analysis of potential departure from operating norms;
- understanding of requirements of EMS standards (where applicable).

Skills

- ability to write clearly and concisely;
- ability to analyse documentary requirements clearly.

💡 Context

Management systems have had a bad press when it comes to paperwork, but try to make them effective without some and you'll see why it is essential. Documents have a crucial role both in moving information about and in providing information on what has happened in the past. The first factor is obvious; having the right information in the right hands at the right time means the difference between managing any aspect of business effectively and trusting to luck.

The second aspect, providing an important historical context, may not be quite so clear. Why does an organization have to know where it has been as well as where it is going? If you don't keep one eye on what has happened in the past, it can suddenly overtake you; minor departures from limits become major problems, the complex relationships between the different stages of processes cannot be explored, and the causes of problems remain a mystery, doomed to be repeated with resource (and morale) sapping regularity.

In this chapter, we are using the word documentation to mean a written form of procedure, record or instruction, regardless of the medium, whether it is paper or electronic. Many of us may still be waiting for the paperless office; a well-thought through structure of documents can be infinitely more flexible if held on computer disks and networks. Whatever the medium, a key point is to think carefully about what is really needed in the way of overall documentation. Although consultants may have their own ideas and recommendations, in this area, no one knows a company better than its own management.

If you are pursuing ISO 14001, EMAS or a phased implementation scheme, check through the management system specification and carefully make a note of whether the stated requirement is for a 'procedure', or a 'documented procedure'. You may also come across the requirement to 'keep records'. These indicate that a document of some sort will be necessary, whatever your own judgement on the matter. Remember too that document requirements will include non-procedural documents like a description of the EMS scope, the environmental policy, a description of the main EMS elements and their interaction (aka 'a system manual'), records and so on.

Beyond this basic approach, in trying to decide whether documentation is needed or not, make some kind of risk assessment of the likelihood of a problem occurring in its absence. A useful rule of thumb is thus to provide documents only where the absence of them might result in an adverse impact on the environment. For example, if you are using only your own in-house drivers to move and deliver bulk chemicals to the site, once you have trained them in the procedures involved in cleaning, connecting up and so on, and then go on to record this in your training records, do you really need a written work instruction on the topic as well?

The answer to that question is that it depends on the particular chemicals being used, the potential risk of getting it wrong and how big a hazard is caused if something does go wrong. If an external assessor were to come across an undocumented procedure, could you support your decision not to provide or use one? If it is a conscious reckoning of the hazards against the effectiveness of the existing measures, and there is no requirement in the standard specification, then an assessor may well accept that a document isn't necessary (after all documented procedures are not guarantees against accidents). On the other hand, if there has been little or no thought given to the subject, an external assessor will be hard pressed to tell the difference between a managerial oversight and a deliberate decision. The idea is to avoid nasty surprises, and documents (or even just thinking about whether documents are needed in the first place) can help to do this.

Many companies will have had ISO 9000 QMS installed, and may want to look at the way those systems are documented. There are two things to remember here, if this is the case for you. The first is that, in our experience some consultants who have installed these systems have had a regrettable tendency to write documents by the yard (or metric equivalent). Secondly, be wary of how closely you need to reproduce documentation. By all means consider the similarities, such as a ready-made document control system, but cloning is for ethical debate, not management system documentation.

Think about what the document will be used for internally and externally, but don't become blinkered to the point where paperwork is produced as a default position, 'just in case'. It is true that documents are basic carriers of information, but they are not the only way information can be transmitted. Always consider the alternatives carefully: training, existing records, other checking systems such as shift inspections and so on.

Tasks

Will you know when you have written enough procedures?

There are many internal assumptions within this question. The first assumption is that all procedures have to be written, the second that there is a finite amount of documentation that in some way is magically 'right' for each management system and organization, and the last is that managers keep writing until something indicates that they should stop. Let's take them in reverse order.

Instead of writing to meet some kind of undefined target amount of paper, it is better to start from the position of asking 'What documents/procedures do I really need?' This is obviously going to bring you into direct contact with the baseline requirements that already define the EMS (covered in previous chapters), that is, the law, the environmental policy and the objectives and targets that have developed from them. The law may include requirements for you to keep certain documentation and these should obviously be considered 'must haves', and 'must be able to reference easily'. Company policies, objectives and targets will define the areas of environmental management and performance that will need possibly three types of documentation (records, system documents and procedures).

For those not using ISO 14001, EMAS and/or a phased implementation scheme as a guide, think about how you are going to get an overview of the EMS itself as a distinct strand in your existing management system. How do you do it, for, say, health and safety? Or quality? Many people create what they call a 'manual' which provides a central resource that describes all the elements of the system and how they interact with each other. If you have one already for quality, say, then it is easier to build the EMS strand into this manual than it is to create another separate manual.

When an EMS is new, it will be competing for space and time with all the other demands of management and their associated systems. There is always the fear that without visible documentation, the new systems will somehow not make themselves felt, and their requirements thus fall by the wayside. This is not unreasonable in many ways, but to use documentation to allay that fear has a glitch in the logic. It is rather like shouting in meetings so that you won't be ignored. The tactic may work temporarily, but at what cost? The original idea is to impart information in such a way that it is acted upon, and not to draw attention to the way that you have delivered the message.

Do you know how to go about writing procedures?

Think carefully about how useful it might be to have a procedure for writing procedures, and documentation for documents. Although it sounds as though this is a 'formalization too far' as one manager put it to us, there are definite advantages; the main one being that it saves having to rethink procedures every time circumstances change. Note that ISO 14001 and EMAS require that there is a procedure to approve all documents (not just procedures) for adequacy 'prior to use', so for those pursing external recognition, it is a non-optional formality. The basic steps that go towards effective documentation are as follows:

1 Analyse need and define purpose

Documentation and procedures fall into roughly five categories, which can be remembered using the mnemonic 'Documentation always comes in P.A.I.R.S.'

Planning: How do you plan new projects? What about the environmental aspects of research and development, production, purchasing, design, finance and so on?

Activity: How do you control processes, suppliers and contractors, raw material use, pollution abatement, monitoring of emissions and discharges, maintenance routines?

Information: How do you know who goes to what meetings? This might include minutes, agendas, circulation lists and the like. How do you get verification that information has been received?

Risk: What do you need to cover safety issues and emergency drills? Is there anything outlining your response to accidents, or other abnormal contingencies? And what about information for the emergency services?

System: How do you produce and change your procedures? How do you know corrective and preventive actions have been followed up? What about procedures for identifying legal requirements, significant impacts and other vital information?

2 Draft

Once you've identified all the needs and analysed exactly what each procedure is designed to achieve, you can start to draft a procedure. Your previous analysis will also have included what you already do, and how many other existing procedures there are allied to the new one. If you are not already using flow charts to outline procedures, consider introducing them. Whatever format you use, expect your draft to be modified extensively in the following steps.

3 Test

The best suggestion that we have heard here is to try the procedure out twice, once with someone who knows nothing of the actual job to be done and once with an old hand. Each will have their own perspective and each will spot different facets of the procedure that need changing. Having used this method, we can vouch for its usefulness.

4 Revise and print

You'll need to ensure that your document control procedures are among the first ones that you write, because when you issue a procedure, you're going to need to know what issue it is, whether there have been any revisions, what the date of issue was and so on. Again, it is a good idea to state the purpose of the procedure; it not only helps users, but it will fix it in the mind of the procedure's author or editor. Ensure that you can also list responsibilities, a basic description of the procedure, any cross-referencing of other material and/or procedures, and the signature of whoever authorizes it.

5 Updating

Don't forget to review the documentation on a regular basis. You may wish to review each procedure on a 12 month cycle, but think also about how changes in operations in between reviews will trigger an examination of the procedures that apply.

As for documenting your documents, an overview of which documents exist (not just those that relate to the management system itself), where they can be found and who is responsible for them is a handy map to the documented parts of your system. This can be a really useful tool, especially when a revision of one procedure may have a knock-on effect on others, and you need to be able to see the relationships between them. This sort of document can also help eliminate procedures that could end up at odds with one another.

Do you know how to control documents?

All levels of staff may need access to the EMS documentation at some point or other if it is relevant to what they do. How do they know they are looking at the most up-to-date version of that particular document and not something that has been superseded? The answer is exercising control over your documents. In essence, document control means never having to say you're sorry. It answers the basic questions such as:

1. Can the documents be identified?
2. Are they approved?
3. Are they the current versions?
4. What happens to the old and obsolete documentation?
5. Who is responsible for looking after documents?
6. Have they been issued and/or made available to the right people at the right time?

If you are already using an ISO 9000 QMS, your existing document control procedures may well be sufficient, providing that they take account of the extra documentation that will apply to environmental management (ie legal documentation, guidelines and so on).

Do you know which elements of your EMS are affected by documentation and its control?
You may be surprised at how many elements of an EMS are affected by documentation and procedural issues. Consider each element carefully and only create documents that you need, not that you think it might be nice to have.

General requirements
(ISO 14001 Clause 4.1 /EMAS Annex I-A.1)
This clause is so fundamental to the overall management system specification that it normally gets overlooked. Its inclusion in the standard is to ensure that readers understand the difference between those parts of the standard that are guidance (Introduction, definitions, scope, annexes, etc) and those that are the core requirements (the whole of clause 4). Interesting then that a key requirement to include a documented definition of the scope of the EMS is put here, rather than anywhere else. All other elements of the management system are defined by this scope, so make sure it is documented.

Environmental policy
(ISO 14001 Clause 4.2 /EMAS Annex I-A.2)
If you are going for ISO 14001 and EMAS, this has to be documented. If you build it into your health and safety and quality policies, remember that external parties will see all of them, not just the environmental parts. You will also need to ensure that the policy is aligned with the documented EMS scope. Effective distribution to staff is also an important issue here, as awareness of the overall policy by all employees is crucial to a successful EMS.

Objectives, targets and programme(s)
(ISO 14001 Clause 4.3.3 /EMAS Annex I-A.3.3)
Again, if you are looking for ISO 14001 and EMAS registration, these elements have to be documented in such a way as to be accessible to relevant personnel. Your policy and the nature of your objectives define what is relevant.

Resources, roles, responsibility and authority
(ISO 14001 Clause 4.4.1 /EMAS Annex I-A.4.1)
The standards are clear on this issue. The structure and associated responsibilities need to be defined and documented, mainly to avoid the mutual recrimination that happens after something goes wrong. Everyone should know what they have to do, both on a day-to-day basis and when things don't go according to plan. A chart or organogram outlining the structure of responsibilities for environmental work may be sufficient, but make sure the right people have access to it.

Communication
(ISO 14001 Clause 4.4.3 /Annex I-A.4.3)

Interestingly, a documentary requirement that is often overlooked is the one that stipulates organizations should document their decision relating to whether they will communicate with external parties about their significant environmental aspects. Whatever decision is finally made, even if it is a negative one, it should at least be documented somewhere in the system. A good place to start if you want further information as to what might be entailed in EMS related communications is the International Standard *ISO 14063, Environmental Communications*.

Documentation
(ISO 14001 Clause 4.4.4 /Annex I-A.4.4)

The system itself will require a certain amount of documentation in order to be maintained in an efficient manner. If you are pursuing ISO 14001 then these requirements are defined as 'a description of the main elements of the EMS' and their interaction. Be careful as to the strict definition of what constitutes core elements and check with your final choice of certifier (we have seen lists that don't reference 'audits' as a main element, for example).

Many organizations choose to organize the contents of a central 'management system manual' around the clause structure of ISO 14001, to ensure that they have covered all the requirements. If you decide to use the structure of your existing ISO 9000 QMS manual instead, it's probably a good idea to provide a cross-referential matrix, so that you (and your external assessor) can find a way through the documentation quickly and easily. Don't forget to reference other existing documents where necessary, and there's some extra guidance available in both ISO 14004 and ISO 14001's annexes.

Control of documents
(ISO 14001 Clause 4.4.5 /EMAS Annex I-A.4.5)

By documenting the approval and control procedures you can ensure that your document control procedures are themselves consistently followed. If you've got an ISO 9000 system already, don't bother to re-invent the wheel, just use the same procedures, but remember to adjust them to take account of legally related documentation (permits, licences, consents, etc). If you haven't got document controls already in existence, ISO 14001 is pretty clear about the requirements, though take special care about the removal of obsolete documents. Environmental legislation is very fluid and requirements can change almost overnight, so using an obsolete document with the wrong information on it could prove to be disastrous. Remember, too that there may be a certain amount of externally originated documentation that you need to run your EMS. This will need to be identified and controlled in the same manner as your internal documents.

Operational control
(ISO 14001 Clause 4.4.6 /EMAS Annex I-A.4.6)

This relates specifically to written procedures, which may well form the skeleton of your operational controls. Training and other types of procedures will put the flesh on the bones,

DOING

but you should tie the written procedures directly to operations or activities that in turn affect the significant environmental impacts you have already identified. The reasoning is that because the impact is affected by x number of activities, it only takes one of those activities to be carried out in a way which does not complement others in the EMS structure, and the impact could slide out of control. Ask yourself, how many of those activities are affected by procedures and how many of those procedures actually need documenting? How many are already covered by training or other means? How does someone know that they've done something right when they've done it? How do my suppliers and contractors know what they have to do in order for us to manage our significant impacts? Is there potential for the absence of a documented procedure to lead to a deviation from our policy, objectives and targets? The answers will indicate what procedures you need and whether they need to be documented.

Monitoring and measurement
(ISO 14001 Clause 4.5.1 /EMAS Annex I-A.5.1)

ISO 14001 is direct about this area, asking for documented procedures covering the monitoring and measurement of 'key characteristics' which are related to the previously identified significant impacts. Some monitoring is required by law, so as a minimum, make sure your procedures match those requirements and that you have some way of auditing against those requirements (ie if a regulator instructs you to monitor an outflow on a daily basis, and your procedures ask for monitoring on a weekly basis, how would you discover that such a problem existed?). Don't forget the monitoring equipment itself; this will have to be calibrated and maintained regularly, so ensure that procedures and records are clear on these issues.

Evaluation of compliance
(ISO 14001 Clause 4.5.2 /EMAS Annex I-A.5.2)

Although there is no requirement under this clause for a documented procedure, there is a requirement to keep records of the periodic evaluations of legal compliance and compliance with identified 'other requirements' that you carry out. You will probably find it useful to link these records to your corrective and preventive action records so that audit trails are easy to follow (see below). You may also want to use the records as part of the management review process.

Non-conformity, corrective action and preventive action
(ISO 14001 Clause 4.5.3 /EMAS Annex I-A.5.3)

It is difficult to undertake effective corrective and preventive actions without some form of documentation, especially if the changes required affect the procedures themselves. Aim to leave a neat, cross-referential paper trail so that identified problems can be traced from being noticed as part of the internal audit through to the action that is taken to correct them. That way, the system is kept vibrant; alive to changing circumstances, processes and materials. It also means that no one can simply forget when it comes to putting things right.

Control of Records
(ISO 14001 Clause 4.5.4 /EMAS Annex I-A.5.4)

The law will require that you keep defined records, and maintain them over a specific amount of time; you will need to check this. In addition, your procedures for record keeping should cover training and audit records and any other type of evidence that relates to the way you are managing your environmental impacts (ie legal and other requirements, communications with external parties, objectives and targets, calibration records, monitoring and inspection data, non-conformities, corrective actions and management reviews).

Management review
(ISO 14001 clause 4.6 /EMAS Annex I-A.6)

The management review is the time when senior management of the organization get to take a cool, reflective look at the way the management system is working and whether it is still synchronized with the other business drivers of the organization and what opportunities for improvement exist. Conscious decisions need to be made, rather than having issues swept under the table and scapegoats conveniently set up for later sacrifice. The management review is part of this; documenting the review provides records of the decisions made, which may prove valuable when circumstances change in the future.

What are the extra documentation requirements in EMAS?

While EMAS does not make quite such play of using documentation in its wording, it would be hard to provide full evidence required by an EMAS verifier without some form of documentation. The scheme is also closely aligned with the requirements of ISO 14001, thus to all intents and purposes, the management system requirements in this area are very similar. However, there may be extra documentary requirements in EMAS which are extremely important, namely the provision of objective evidence that a dialogue with 'interested parties' has not only been established, but is also ongoing. Environmental statements, the raison d'être of EMAS, have to be validated annually (unless yours is a small organization) so there will obviously be more documentation relating to these as well.

 Narratives

Smallco

The management team, including the MD, agree that documents are needed if the EMS is to function effectively. The Ops Manager has clearly identified what documentation is needed, having carefully interpreted this chapter in the context of Smallco's needs. There is significant shop-floor resistance to 'more bloody paperwork'. Again, there are some communication issues here, which the Ops Manager can in part address in the communication briefings noted in Chapter 9. However, there is also a critical requirement that accountable staff do keep the new EMS documentation for which they are responsible up to date, and in ways that ensure its integrity. These documents, after all, form part of the objective evidence that the EMS is doing what it says it will do.

The Ops Manager gets authorization from the management team to establish a series of working groups to draft new procedures to deal with EMS documentation and document control. He recruits from each workplace to ensure representation and works with these groups to establish meaningful procedures which staff can own rather than imposing his own on an unsuspecting workforce. These drafts are taken back to workplace meetings and tested by the representative staff before being endorsed by both these meetings and the MD.

BIG Inc

Document control of BIG Inc's ISO 9000 system, even though handled by a sophisticated in-house electronic manager, seemed to produce reams of paper. It looks like the EMS is set to repeat that pattern, though no one is quite sure why, especially when the project manager has taken such trouble to ensure that only the minimum number of new procedures and work instructions has been developed.

Diagnosing the situation, the IT Director suggests that the problem is not the origination of the material but the control of the document database itself. The EMR agrees with this to an extent, but on examining the problem more closely and talking to the users of the procedures, he discovers that many of the staff find the procedures themselves too complex to be easily understood, and print them off because they often need to be referred to while their computers are already engaged in a series of complex tasks.

The changes asked for by the EMR are thus twofold. The first is that all procedures for the EMS are written as process flow charts and secondly that the charts are included in the 'knowledge management' application that is already a part of the company network. This application, as well as having restricted access controls, can be called up as a separate 'window' on all computers, even while multi-tasking, removing the need to print off material. Control of the database then rests with EMR and the project manager who is the only source of printed material. This system design is also incorporated in customization projects for BIG Inc's clients adding value to the organization's existing products.

Trade secrets

- Remember that procedures can be implemented, but still may not be effective. How are you going to measure the success of a procedure?
- If a procedure is not being implemented, don't automatically blame staff. Use that information as valuable feedback and ask for more.
- The structure of your documentation is going to reflect the nature and range of operations that your organization carries out. Adapt existing 'maps' where possible.
- The documentation of a QMS is much more limited in scope than that required for environmental work. If using or adapting a QMS, check that you have enough 'coverage'.
- By all means meet the requirements of a specific standard, if you are using it, but don't be afraid to simplify your existing documentation if your analysis indicates it.

- Where are your most significant impacts or potential impacts? Write the procedures to deal with these first.

⁇ Things to think about

- Ever since EMSs were first developed, regulators have been attempting to assess their relative value in terms of addressing legal compliance. Regulators worldwide suffer from limited resources, and if a link can be proven between high levels of compliance and the existence of an EMS within an organization, such resources can be allocated on a risk-based approach (ie putting it crudely, those organizations with an EMS do not get inspected as frequently as those who do not have one). Such a link has yet to be established, but that is not to say that it is beyond the bounds of possibility. Either way, making the evaluation of legal compliance a thorough and possibly discrete exercise may become increasingly important.
- Increasingly, organizations are linking their environmental documentation to their financial systems. Not only does this lead to a greater alignment of quantifiable business and environmental benefits, it also brings the ability to manage an organization towards sustainability that much closer.

DOING

12

Operational Controls: Can We Do This?

Operational controls link intention to reality in the same way that nerves and senses link thoughts and the physical world. We can probably all think of organizations who are either 'all brain' (great product design, shame about the quality) or 'all brawn' (terrific marketing and after sales, shame about the product). These are extremes which could be ameliorated by better overall control between intention and action. When you are managing environmental impacts, you'll need to make changes at this level.

✓ ISO/EMAS quick check

Area of EMS	ISO 14001	EMS
Operational control	4.4.6	Annex I-A.4.6
Emergency preparedness and response	4.4.7	Annex I-A.4.7
Environmental aspects	4.3.1	Annex I-A.3.1
Legal and other requirements	4.3.2	Annex I-A.3.2

☐ Chapter executive summary

Move to the next chapter when you can answer all of the following questions in ways which make sense to you and your organization:

How do you know that you've got the right controls?
How do you know that the controls will work under all conditions?
How do you manage control issues that affect your suppliers and contractors?
How are emergencies and contingencies addressed?

⌖ You are here

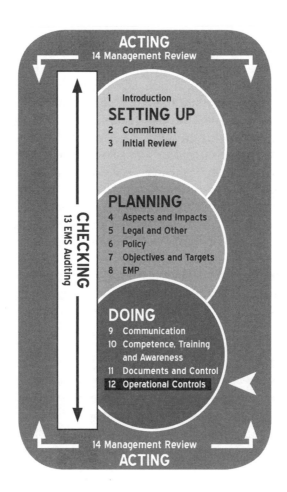

🔧 Toolkit requirements

To work through this chapter you will need the following knowledge and skills:

Knowledge

- range of your organization's operational controls relative to the significant environmental impacts;
- risk analysis of potential and actual environmental hazards;
- understanding of requirements of EMS standards (where applicable).

Skills

- ability to analyse number and type of controls needed;
- ability to assess risk under a variety of different circumstances.

Context

In your own organization, many of the operational controls will either be based on 'common sense' or have evolved over a long period of time. It is possible that no one actually notices how they work any more, and certainly when changes are made to incorporate new ideas like environmental impacts, not only the controls, but their interaction has a tendency to change and produce further anomalies. In short, it is time to look long and hard at how you do what you do, and what you do when things go wrong.

Many organizations find that ensuring they have sufficient operational control is the most challenging part of installing a formal management system, whether they are tackling quality, health and safety or environmental work. It is the level of cool detachment required which many managers find the most difficult thing to achieve. Even when this sense of distance is successfully acquired, it is hard to look at why you do what you do and think of new ways of doing it. It is around now that environmental managers start to hear the cry of 'But we've always done it this way', as though that were reason enough not to change.

However, the more detailed the analysis of the operations, the easier it is to make changes to achieve improvement not only in environmental performance, but perhaps in other areas as well. Much of this work may already have been done in the IER (see Chapter 3). The analysis should consider how controls affect the identified significant impacts, and how they are used during emergencies, unusual operating criteria, and especially when suppliers and contractors are involved.

As an environmental manager once opined to us on this subject, 'If the steering wheel isn't connected to the road wheels, it's going to be a short drive'. When intention is to be carried through to reality, the relevant operational controls must be in place and work together to ensure not only legislative compliance, but also conformity with the organization's own stated objectives and targets. It doesn't matter how many people are on board if no one can drive.

Tasks

How do you know that you've got the right controls?

The one way to ensure this is to focus on exactly what it is that you are trying to control or manage. The IER (see Chapter 3) will have revealed to you what are potentially or actually the most significant environmental impacts of your organization. You'll have built these into your policy, objectives, targets and management plan. Now is the time to ensure that those responsible for the plan have the controls at their disposal to meet the objectives.

For example, if your most significant environmental impact is a particular water discharge, your policy will be to prevent pollution, stay inside the law and have a plan in place to reduce the harmful content of the discharge even further than the law requires. If you look at that in detail, the controls you will need are:

- emergency plans and procedures (prevention of pollution);
- controls that relate to the legal requirements (monitoring equipment, process controls); and
- controls that have a bearing on the discharge content (process controls).

Look for inconsistencies by following a hypothetical audit trail. Choose any of your objectives and for each one:

1   identify the environmental impact;
2   identify the activities that affect or give rise to that impact (include the activities of contractors and suppliers);
3   identify the controls that are exercised over those activities;
4   identify where control is not sufficient (lack of implementation);
5   decide whether the operating criteria cited in the controls are sufficient (lack of effectiveness);
6   repeat the steps above twice more for the same impact, and this time consider abnormal and emergency conditions.

In order to get full coverage, it may also help to think of your organization's range of activities in a variety of ways:

•   short term (day-to-day management, focused on the implementation of existing controls);
•   medium term (activities relating to pollution prevention, conservation of resources, changes in processes and design management, focused on the effectiveness of the controls);
•   long term (strategic business planning, focused on establishing the organization's intent and policy).

This long-term element is the most frequently overlooked aspect of the work, mainly because boards and senior management involved in planning regard environmental impacts and their operational controls as something to be considered after the fact rather than part of the planning process. The result is a lot of post hoc rationalization that maintains the planning status quo without allowing for the missed opportunities.

How do you know that the controls will work under a range of conditions?

This will very much depend on the depth of the analysis that you go through when you run your imaginary audit trail (see above). The key to developing successful controls is the identification of the activities associated with the impact you are attempting to manage. Let's take the example of the water discharge again. The obvious place to look for controls is all the way through the process that gives rise to the discharge in the first place. But for really effective control, a manager has to throw the net wider in order to catch the whole range of factors that may have a bearing on the quality of the discharge. Consider the following, for instance:

•   What effect do research and development or product design have?
•   Could you change your purchasing criteria?
•   Could you change your raw materials?

- Are there any problems caused by maintenance?
- How are the aspects that are contracted out handled?
- Could your current storage and handling routines be contributing?
- What about support functions, do they make a contribution?
- Are there any related impacts or activities that take place away from your site?

You will need to look at all your relevant operations and activities in different states (normal conditions, abnormal conditions and emergencies) and over different times (past, present and future) to make sure that you have considered the full range of activities and associated operating conditions and developed controls that cover each circumstance. If you aren't sure exactly what an 'abnormal' condition is, then think about unusual activities that take place at prolonged intervals, such as maintenance (which is or should be far from a continual process), starting up or shutting down a process (which may give rise to unusual peaks or troughs in outputs), or even unplanned occurrences like rainstorms (which may require management of excess water). If you are still unsure, take another look at Table 4.4.

The reason for looking at past activities is obvious when it comes to issues like contaminated land, especially as what has happened on the site or sites in question may have a bearing on future development plans. 'Brownfield' sites in particular can be problematic, especially as previous environmental regulations were not as strict as they are now. Nineteenth and early 20th century industrial processes were notoriously toxic, and their disposal routes are not always recorded. These sorts of issues could affect your future use of the land, and perhaps even your ability to sell it on to others without incurring high remediation costs.

How do you manage control issues that affect your suppliers and contractors?

You may find that some of the activities that have a direct bearing on your significant impacts are not directly in your control. Those who supply raw materials or sub assemblies may carry out their work in such a way that subsequent activities on your site(s) are affected. Contractors who you may ask on to your site in order to carry out specialized work may similarly affect some of your outputs. It may be a boiler maintenance crew, or building work, or waste disposal, or even warehousing and transportation. If this is the case, you will still need to exert some kind of control or influence over such activities, if you wish to maintain your environmental performance objectives intact.

Although it would not make sense for you to try and control suppliers and contractors directly, as a customer or client you are still in a position of considerable influence, and your EMS should be set up to use this in order to meet your policy commitments. There will be a point where your management system stops and your contractor's or supplier's will start, and it is important that this interface is acknowledged and managed effectively. For example, if the contractors in question are regularly on the site, is it worth considering including them in the environmentally-related training that you will be giving your own staff? At the very least you will need to take steps to make sure that contractors know what your policy towards the environment is, and what they have to do in order to meet it. In which case, it makes sense to utilize existing systems to maximum effect and include them in the training.

This may take some close discussion with your contractors, as they may not be aware of which of their activities impinges on your environmental performance. Disposal of leftover material from contracted work is the most usual problem, especially when it comes to getting rid of solvents, paints and associated contaminated material. You may have many contractors or subcontractors to deal with at once, especially if you are involved in building and construction, so the need to exert an influence and come to an understanding as to defined responsibilities is even more important here.

Suppliers can be a different matter. Many managers discuss issues related to waste packaging, design and material handling with their suppliers. Many more feel reluctant to approach their suppliers about what they are trying to do, especially when the supplier dominates the market or is many times bigger than the customer. We recommend always approaching suppliers, however, as many managers have been surprised by the positive response with which they have been met. For example, when an electrical subcomponent manufacturer talked to their suppliers about reducing the amount of packaging used in the transfer of materials, a small team was formed which included representatives of all the companies concerned. The team decided to research alternative methods of delivery and reusable packaging, and after six months came up with a series of suggestions that involved new methods for both the suppliers and their clients, as well as new materials. When implemented, these suggestions cut the total weight of packing disposed of by over 50 per cent, and had a measurable impact on lowering the cost of deliveries as well.

The LCA of your products and activities, which you will have carried out as part of your IER, will certainly help to reveal how your organization stands in relation to the rest of the supply chain. Many supplying companies already appreciate that their customers are linked to them and vice versa, especially when it comes to reducing adverse environmental impacts. As a result, they may well be willing to explore issues with you and consider alternative approaches to their current practice; many already appreciate that this type of collaborative approach could possibly net them financial and other business benefits as well.

If you haven't done already, now is the time to get the procurement function of your organization switched on to the possibilities of environmental management and its associated drivers. Procurement departments are a veritable mine of hard won information concerning the nature of your supply chain. They may never have looked at that information from the perspective of environmental impacts though and may not realize that they already have valuable data that can be used to drive resource efficiency. However, once they are persuaded that there are direct benefits to their function, there can be no greater champion for environmental improvement than a converted buying manager! Similarly, the teams responsible for new contracts and tendering will be useful, once they realize that they might be able to use the EMS and its operational controls to exceed requirements of new customers, not just meet them.

Ultimately, with both contractors and suppliers, you are in the position of being able to take your custom elsewhere, and no one who wants to stay in business likes to lose

DOING

a customer, no matter what their size. However, we also know that many companies are loath to move their custom in this way, owing to the upheaval of developing a new working relationship with a new contractor or supplier. No one would deny that 'breaking in' newcomers to the supply or tender lists can be time-consuming, but if your current suppliers and contractors are not prepared to work with you, then you are putting at risk your own company's commitment to environmental improvement. The line of least resistance could turn out to be the most expensive one in these circumstances.

Where you have a sole supplier and a sole contractor, the position will need very careful handling. Again, a collaborative approach bears the best fruit.

How are emergencies and contingencies taken into account?

Some of the controls and procedures that we are talking about in this chapter (and the previous one) will relate to emergencies, and the responses to them. There will need to be emergency plans and drills, which in turn are practised to ensure that they work. (A real life example was given to us of a gateman who followed his instructions to the letter and did not allow anyone on to the site once the alarm had sounded. Unfortunately, he interpreted this as meaning that he shouldn't allow the fire service on to the site either. Suddenly, drills start to make a lot of sense).

You may have emergency plans that already cover fire and accidental hazards created by or related to chemicals, but don't assume that they will provide enough coverage for your environmental policy. It is worth looking again, because such plans may only relate to the health and safety aspects of the emergency, not the environmental aspects of them. For instance, if a fire breaks out in a store of hazardous material, existing drills are probably focused on removing personnel and anyone in a nearby residential area from immediate harm; but has any consideration been given to air emissions, fire water run off (which can affect land as well as water) and other impacts on the environment?

You should also remember that some of your potential environmental impacts (identified during your IER in Chapter 3) may not be included in your existing emergency procedures for the same reason. The risk assessment which you have used to identify significant environmental impacts will also indicate where your emergency and contingency work should be applied. Interestingly, when you have undertaken this review of your existing emergency procedures, it may also be worth double checking your insurance policies. You may find that the cost of remediating an environmental impact associated with an emergency is not covered. This can certainly help focus the mind on the true cost of accidents.

 Narratives

Smallco

There is a danger that the term 'operational controls' could be misunderstood by the work-force as yet another set of management devices to restrict working practices. Clearly this is not the case, which needs to be communicated effectively to the people on the ground who will ultimately decide just how and if the EMS functions properly. From his work on aspects and impacts, the Ops Manager decides to integrate a series of workshops into his briefing sessions about the EMS itself. The aspects and impacts work has provided 87 examples of issues that need tackling, and a good spread through Smallco's operations. The workshops take two or three pertinent examples of environmental impacts and challenge the participants to devise strategies to manage the aspects of their activities to reduce these impacts. Clever use is made of the checklist in ISO 14004, which provides practical help on operational control. The language used in the workshops is not the jargon of EMS, but relevant to the participants.

The workshops are a great success, so much so that ideas keep coming up in other meetings when environment is on the agenda and through direct approaches to the Ops Manager. The Ops Manager uses these ideas to construct workable draft operational controls and procedures which are played back to the workplace groups before being finalized. Although this way of doing things is more work in the short term, the Ops Manager believes getting it right for the EMS and the workforce will bring medium-term benefits such as not having to fire-fight or undo unworkable control systems. In fact, word has got about that Smallco are running these workshops and the Ops Manager has been invited to speak at a regional environmental event and run some workshops for other companies. Whether he can find space to do this is another story, but the positive messages from outside do much to reinforce his reputation within Smallco.

Smallco have now reached the end of Phase 4 of their phased implementation and are ready to develop their EMS audit programme.

BIG Inc

At first, there is much concern within the company that 'operational controls' in terms of influencing client decisions will be hard to identify, let alone install. As it happens, the job turns out to be much more straightforward than anyone thought, mainly because of the Head of Consultancy, who had already clearly defined the process controls for the design and consultancy functions for inclusion on the knowledge management database.

More work turns out to be needed from the project manager in terms of ensuring that the operational controls concerning energy management are correctly installed and used. Though there are climate control devices in the headquarters building, these are being abused owing to staff not understanding their proper function (including a member of staff who uses a thermostat to brighten the lights over his workstation, mistaking it for a light dimmer). Full training is provided to all staff, including new recruits, on all relevant aspects of energy management.

Other areas where operational controls may not be working properly concern the buying function, where new procedures were not being followed, transport, where staff and associates were not completely aware of new policies, and emergency procedures covering electrical fires within the office buildings.

Trade secrets

- Emergencies need more than a procedure; they need an attitude of mind. Think about making it everyone's responsibility to report anything unusual.

- Be aware that health and safety management only looks after the workplace environment and the staff who are in it, not environmental impacts which may be non-personnel related.

- Compare your controls to your objectives; do they match? If they do, are they enough?

- Don't assume that having the technology means being able to use it. Untrained operators won't be able to follow procedures, no matter how well written.

- Create consistency. Draw a line from your significant aspects and impacts through the policy, through your objectives and targets, and through your management plan to the controls. Do they all line up in your sights? Are there any holes?

Things to think about

- Technology can be an unpredictable element in terms of exerting operational control. There are times when next generation equipment simply apes existing methods and techniques, but does it continuously or faster. There are other occasions when new technology makes some procedures obsolete, creating a knock-on effect throughout management processes by doing things in an entirely different way from what went before. As technologies become more specialized, there is also the increasing likelihood of them competing instead of working together within an organization. Be aware of potential 'system fallout' whenever new technology is introduced.

- Phased implementation schemes place such an important role on operational controls that they prioritize them before other formal elements of an EMS. They usually occur at a midpoint in the total EMS development cycle, well before other aspects of the system such as training, communication and even auditing. The continuing emphasis on environmental regulatory development throughout Europe and internationally accounts for this in part.

- Specific environmental contractual requirements are becoming increasingly popular with all types of customers, and in the future, it is more likely that these will be backed by specific inspections now that phased implementation schemes can make such an exercise relatively easy. Winning and retaining new business will be greatly helped by the demonstration of good operational control.

Checking

13

Environmental Management System Auditing: Are We Doing What We Say?

Auditing is an organization's nervous system, constantly taking messages and sending re-adjustments to and from all areas of the system's body. The more sensitive and responsive the system, the better it is. An effective audit enables the organization to keep 'on track' of its original objectives as well as checking conformity against the management system elements. A good audit methodology can help to identify the types of problems that occur and whether they are to be solved by attention in the areas of 'Direction', 'Realization' or 'Delivery'.

✓ ISO/EMAS quick check

Area of EMS	ISO 14001	EMAS
Evaluation of compliance	4.5.2	Annex I-A.5.2
Internal audit	4.5.5	Annex I-A.5.5
Non-conformity, corrective action and preventive action	4.5.3	Annex I-A.5.3
		Annex III

▢ Chapter executive summary

Move to the next chapter when you can answer all of the following questions in ways which make sense to you and your organization:

How can I structure the auditing process?
What about ISO 19011 ?
Initiating: How are the ground rules established?
Preparing: How does an internal EMS audit plan work?
Conducting: How is an internal EMS audit implemented?
Reporting: How can corrective and preventive actions be truly effective?
What are the key skills to look for in an auditor?
What are the problems of integrating quality, health and safety and environmental audits?
What are the extra EMAS requirements?

🖏 **You are here**

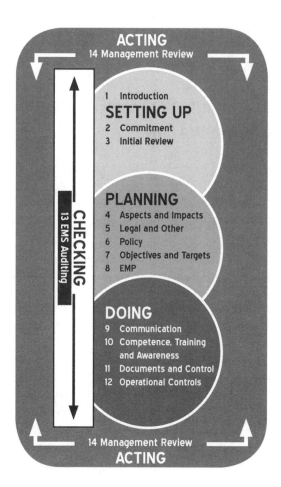

🔧 **Toolkit requirements**

To work through this chapter you will need the following knowledge and skills:

Knowledge

- industry sector;
- environmental impacts and types of control in that sector;
- management systems;
- auditing protocols;
- relevant regulatory and 'other' requirements.

Skills

- communication;
- interpersonal;
- analytical;
- ability to remain objective.

Context

Since almost everyone in an organization either has been or will be part of an internal audit of a management system, this is the highest profile element in the whole EMS. As a result, the perception that most employees will have of a system will be based on their experience of the auditing procedures in general and your audit teams in particular. If this one sobering thought alone makes you review your choice of auditor and auditor training with extra care, then this book will certainly have been worthwhile.

It's down to the organization to state its own purpose when it comes to auditing, and most tend to follow the guidance in ISO 14004 where the standard recommends that audits 'should be conducted at planned intervals to determine and provide information to management on whether the system conforms to planned arrangements and has been properly implemented and maintained.' Even so, within this definition it is possible to put the emphasis more on compliance with the law and the management system requirements, than on the principle of continual improvement. North America was the cradle of environmental 'compliance' auditing as it is known, where the focus is on checking that regulations are being followed. In the UK and the rest of Europe the emphasis tends to be more on improving the organization's performance beyond the law, even though there is a requirement to make a 'periodic evaluation of compliance' in ISO 14001 and, thus, EMAS. We make a basic assumption in this book that people having read as far as Chapter 13 not only want to assure themselves that their organization is compliant, but want to do more than just obey the law.

Whatever the focus, the potential effect of the audit on the whole EMS needs to be understood. We're frequently asked if we think that good auditing will 'save' a bad management system. This may be the case over a period of time, however, how many breaches of regulations, pollution incidents and wasted resources can an organization afford before the audits start to be really effective? In any case, nothing will save an inherently flawed system which has managed to misidentify the impacts it should be attempting to control. The best auditing system in the world may not be able to identify its own 'blind spots' quickly enough.

On the whole, we find that auditing gets a bad name through bad auditors, not because it is in itself inherently flawed. It's a recurring nightmare for those who have been the subject of any kind of management system audit that they will get trapped in a small room with a pedantic auditor. For those who may not have suffered at the hands of one of these inexpertly trained auditors, try to imagine how you might feel if you were told that a particular document 'did not carry the issue number' even though it was clearly identified by the date on its cover, and that this therefore constituted a non-conformity within the management system. This is the sort of audit finding that has simply not been put into the context of the company practice. Such a finding, though technically correct, is functionally useless and undermines the whole process.

At its best, auditing, whether internal or external, can be one of the main contributors to continual improvement, devoid of bureaucratic excesses, giving valuable feedback in order to maximize the benefits of having a management system in the first place. To achieve this, however, organizations and environmental managers have to go beyond the letter of the requirements in standards like ISO 14001 and EMAS and seek out their spirit.

CHECKING

✑ Tasks

How can I structure the auditing process?

For anyone who has never been involved in auditing management systems before, the whole process can seem a bit daunting. The terminology used by professional auditors only helps to increase the anxiety factor, so let's start with a reassuring thought: If you understand the 'why', you can make up your own 'how'. To expand on that, we've found that if people understand what the motives are for a particular action, then they can adapt that action to fit in with their own working practices. If they don't have that understanding, then they tend to fall back on painstaking formality, copying every last detail of an auditing methodology, and 'cloning' what they read in guidance notes and books. Audits designed like this simply miss the point. They don't provide the much sought after feedback the organization needs, the EMS atrophies (as do the staff working on it) and performance levels either remain the same or even get worse.

This section is all about understanding the 'why', and below you can start to refine the how. Let's look at first principles. Being an auditor is very akin to being a detective: both need to gather evidence and stick to the facts. Both will use similar skills to collect the evidence, analyse what they've collected and come to a conclusion about the overall meaning of the facts before them. Collecting and analysing such evidence is required for the maintenance and improvement of the management system and its outputs. A more fundamental way of looking at it is to ensure that your EMS is not carved into tablets of stone, but stays flexible, practical and in synchronization with the organization itself. Without accurate feedback, systems stagnate, and become gradually less and less relevant to the companies they are supposed to be serving.

A successful audit is thus marked by four sets of activities:

1   overall management of the audit process;
2   the collection of appropriate, relevant evidence;
3   the analysis of the meaning of that evidence; and
4   the translation of that meaning into action which will ultimately improve the systems, and the company performance.

Structuring those activities into an EMS auditing system itself is a different kettle of fish altogether, which is where another standard which focuses specifically on management system auditing procedures (ISO 19011) can provide useful guidance.

What about ISO 19011 ?

The full title of the standard is, *ISO 19011 Guidelines for Quality and/or Environmental management system auditing*. Being a guideline and not a specification, it has an advisory role only and, as such, no organization can be certified against it or have a non-conformity that cites it as a requirement. That aside, there's a lot of useful information that can help you to shape your auditing process to make it effective and efficient. The standard splits the process into four main stages:

1   establishing the audit programmes;
2   implementing the programmes;
3   monitoring and reviewing the programmes;
4   improving the programmes.

There are also sections that go into more detail about the competence and evaluation of auditors, listing requirements in a mixture of personal attributes, generic and specialist skills; all of which have to be backed by independence.

However, our experience has been that many managers find it easier to keep the following roughly analogous structure in their own minds:

- initiating;
- preparing;
- conducting;
- reporting.

We've decided to keep this structure for our outline as it allows us to comment in depth on specific issues relating to environmental and legal compliance issues. Remember, the overview we offer is easily adaptable to your own organization's (or ISO 19011's) terminology if required. Comparing either of these lists to the activities previously mentioned, it's possible to use them to ensure that you have a usable blueprint for all your auditing operations. Even better, the same basic structure will apply to internal (first party), customer (second party) and external (third party) audits, so it is worth spending time understanding how the process works in practice.

Initiating: How are the ground rules established?

Initiation centres on producing audit scopes, either for an individual exercise or a whole programme. This should not be confused with the 'scope' of an ISO 9000 quality management assessment or audit, in which an entire management system can be refined to apply solely to a single process. Though this can be done in EMS auditing for an individual audit, when looking at the entire EMS, the scope of the audit programmes should combine to cover the whole organization and the impacts that it has identified. Think too of how you might create an appropriate audit cycle, or in other words, how long it will take to audit the entire system and its outputs.

There is a considerable amount of choice when it comes to deciding on structuring an audit programme and the consequent individual audit scopes. Your decision may also be affected by the way any existing audits are already taking place within the organization. It makes sense to use the same scope, providing that at the end of the programme you can ensure that the whole of the EMS has been covered; so concentrate your attention in the first few audits on any gaps between the quality, or health and safety audits and those elements of the EMS not included. Even so, your choice is still quite wide. Are you going to structure your audits:

CHECKING

- by function?
- by site?
- by department?
- by substance or material?
- by process?
- by product?
- by clause of the standard?
- by work procedure?
- by impact?

Special care is needed in making this decision, as the audit scope can easily be confused with the audit's objective, particularly when there is such an intimate relationship between the two. Whether you are more focused on potential improvement, compliance with a formal standard, investigating the working relationships with suppliers and contractors, or all of the above, understanding the true purpose of the audit will help to define how you put together a complete programme of audits for the whole organization.

Be aware, too, of the special issues surrounding the periodic evaluation of legal compliance required by standards like ISO 14001 and EMAS. Some organizations follow the guidance in both ISO 14004 or ISO 19011 and choose to wrap this evaluation into the scope of the internal EMS audits, which means that legal compliance issues are looked at piecemeal and on the same cyclical and sampled basis as the rest of the management system. Although the standard schemes don't directly stipulate that the evaluation is carried out as a separate exercise, there are several factors that indicate that the 'separate event' is a good approach to adopt.

The first is that certifiers or registrars will usually expect a legal compliance evaluation to have been carried out across the whole documented EMS scope of the EMS before they will issue their certificate. On the other hand, they will not expect an entire cycle of internal audits to have been carried out. At the very least, this means that the periodic legal compliance evaluation needs to be a distinct exercise at the very start of the life of the EMS.

Secondly, think hard about what might be lost if the legal compliance evaluation is rolled up into the internal audit cycle. Establishing legal compliance is complex but fundamental to an effective EMS, so having started with a separate exercise there are several advantages to keeping it a discrete event:

- Auditors maintain their impartiality by not being asked to check areas where they already operate or are affected by legal compliance procedures (how, for instance, could they audit that the evaluation of compliance had been carried out according to company procedures?).
- Technically speaking, legal compliance is an output of the management system, not part of the system, so a system audit may not be sufficient to check such outputs thoroughly.
- Avoiding scoping problems - audits are based on sampling techniques, but the evaluation of compliance has to cover the complete scope of the relevant regulations.

- Compliance issues are frequently interrelated – focusing on them alone will help to maintain clarity when it comes to deciding appropriate corrective of preventive actions.
- Compliance oriented and specifically trained auditors can concentrate on tightly defined issues and be less likely to 'overlook' a compliance issue due to time pressures.
- Rectification of any problems discovered can be prioritized.
- Regulators have easy access to demonstrable evidence of commitment to compliance.

Think too about the risks involved in terms of the nature of the impacts and the risks of not complying with the law. Assessing this risk will help you set out how often you check overall compliance – simple processes may only need checking annually. On the other hand, unless there are very tight controls exercised (such as continuous automated monitoring, itself a form of periodic evaluation), some more complex and therefore higher risk processes might, initially at least, need checking much more frequently. Having established the 'period' in your 'periodic evaluation, remember:

1   The evaluation can include 'other requirements' – if it doesn't, ensure that these appear in the internal audit scopes and overall programme.
2   The evaluation should be subject to an internal audit itself.
3   To make the exercise rigorous, compliance auditors need to be adequately trained in regulatory issues.

Though the rest of this section of the book uses the term 'audit', it is worth remembering that the comments can equally well be applied to a 'legal compliance evaluation' exercise when it is carried out separately.

Once you have defined the scope and objectives of an individual audit, it is then a good idea to ensure that the client, or subject of the audit, agrees with and understands the same definition. After all, if the auditor and the auditee do not agree on what is included in the scope, this may lead to complications in the reporting and corrective action phases of the audits. Remember too, that the agreed scope will be your starting point in deciding how much resource is required in terms of manpower, time and other related areas, such as the frequency of the audit.

When that has been worked through, it is a good time to undertake a review of the relevant documentation that applies to the audited functions. These may be systems documents, legislative or work procedures, but if vital documents are not present or accessible, not only is it a major non-conformity within the system itself, there will also be little point in attempting to conduct the rest of the audit until such a major flaw is rectified.

Preparing: How does an internal EMS audit plan work?

Each individual audit needs an audit plan, simply to keep it on track and delivered on time effectively. How formal you wish to make this plan, and whether you establish a kind of formatted document is really up to you. One thing we would say here, is that there are so

CHECKING

many elements involved in an audit that a reference plan can save a great deal of confusion, rancour and argument as well as adding a positive focus to the activity. Again, one purpose of the plan should be to avoid any nasty surprises for anyone involved in the audit. Obviously no one likes non-conformities being found against their area of operation, but establishing the right open approach at an early stage can contribute greatly to preparing the ground and circumvent overt 'defensiveness' to the audit findings.

Being able to answer a clear list of questions will give managers a good idea of what could be included in their audit plan, so if appropriate:

- What are the audit objectives and scope?
- What criteria will apply during the audit?
- What's going to be audited?
- How does what is going to be audited relate to the EMS?
- Which of those EMS elements are a high priority to examine?
- What are the auditing procedures that will apply?
- What are the reporting languages of the audit?
- Are there any reference documents?
- How long will the audit take (estimate)?
- What are the date, time and place of audit?
- Who is in the audit team?
- What meetings have already been agreed as part of the audit?
- What are the arrangements concerning confidentiality?
- What will the audit report look like and who will it go to?
- Who needs to keep what documents?

There is little point in working through such a list without sharing it with the auditee. It is at this point that any objections raised to the arrangements can be ironed out. This forestalls any problematic snarl-ups which might lead to an audit being aborted. For the same reason, some auditors also like to use a specially formatted checklist, constructed around the scope of the audit to be undertaken, which helps to focus the attention of the auditee on the types of areas going to be examined. There is much to be gained from this approach, though we have heard of objections on the grounds that the auditee is prepared for the audit and may not give a true reflection of the day-to-day working practice. However, we have never known a pre-audit checklist to 'hide' a non-conformity, so in the end it may simply be a matter of taste and company practice.

Conducting: How is an internal EMS audit implemented?

Depending on your organization's culture and size, you may want to formalize the audits with separate 'opening' and 'closing' meetings. These can sometimes be handy, even in a small company, where the detachment of a formal process can take some of the personal 'sting' out of any subsequent findings in the audit report. This is a judgement call for the manager involved in the individual circumstances of the audit, but if you choose depersonalization, make sure it is consistently applied.

The purpose of such meetings is to provide reference points in the process, and gives the auditees the chance to understand exactly what is going on and why. It is another good way of checking that everyone shares the same understanding of auditing and avoids nasty surprises for all concerned. Problems with logistics, availability of personnel, changed circumstances and other factors that may affect the audit can be discussed and action agreed.

At the end of the process, the closing meeting helps to 'sign off' the individual findings of the audit, and gets the auditee to accept responsibility for the subsequent corrective (or preventive) action. It is possible that without this formal touch, the 'patients' may stay in denial and the problems fail to be rectified. If you are pursuing formal standards, ISO 14001 requires that responsibilities for handling corrective actions and any further investigation associated with them should be made clear through a proper procedure and that the actions themselves should be appropriate to the scale and nature of the problem. A documented closing meeting can go some way to meeting this requirement.

The easiest way to describe the audit process is to differentiate between the ways of collecting evidence. The three main methods to collect objective evidence are:

- interviewing: conducting structured interviews with selected personnel;
- viewing: watching actual practice and operations taking place; and
- reviewing: sampling documentation for internal and external consistency.

Interviewing relates to individual staff members at all levels within the scope of the audit, selected either randomly or on the basis of their responsibility. Communication and interpersonal skills are particularly important to get the most evidence out of interviews (see auditor skills below). It is here that a good auditor can establish the general level of comprehension that underpins the EMS. Although environmental management may be an entire job for the environmental manager, and even for the auditors concerned, to everyone else it is merely another factor to be included along with all the rest that make up day-to-day operations in an organization. Interviews can also help to point the way to other potential 'audit trails' that will lead to problems, and at the very least provide a spectrum of perceptions around the particular area being audited.

Viewing, on the other hand, is the simple act of staying alert and noticing details of operations currently taking place, as well as past and potential problems. Sometimes the evidence may be visual, but it could equally well be aural or olfactory. (In training EMS auditors, we used to say 'use all five of your senses', until one delegate questioned whether it would be such a good idea to lick the solvent off something!) One of the main differences between QMS auditing and the environmental type is the sheer breadth of potential evidence to collect, and the fact that there is much more emphasis on its physical nature. This can sometimes mislead auditors into worrying more about the effect on the environment than on discovering why the management system allowed the impact to happen. It is worth reinforcing that all audit trails should lead back into the system, telling you something about the management of the impacts, not the change on the environment brought about by the impact.

CHECKING

Finally, **reviewing** is something that happens to documents (including system documentation), records and monitoring data. This type of evidence gathering is particularly handy in getting an overview of current and past operations. It is possible for interviews to go well, and for observable practice to match the apparent level of understanding, but it is amazing how many times inconsistencies in the paperwork turn up much larger problems. Conversely, and depending on the business culture of the organization, it is equally possible to find all the paperwork in order, only for the practice and the interviews to show that what goes on at ground level, doesn't reflect the neat files and records.

The three different activities in the audit are potentially going to bring together a large amount of evidence. This is the basic raw data which can be used to diagnose any apparent problems with the EMS. The evidence itself may be spread across a wide variety of physical locations and make itself known in an equally wide variety of ways. All evidence will fall into one of these five categories (if you find any more, write to us and we'll include them in later editions):

1   Material, for example, stains on the floor, pollution control equipment installation, shop floor practice.
2   Referential, for example, during an interview, it becomes clear that an incident may not have been reported.
3   Inferential, for example, during an interview, the subject does not appear to be certain of their responsibilities.
4   Documentary, for example, inconsistent records, regulatory permits or out-of-date reference material.
5   Verbal, for example, an auditor is told that they should have been there last week, when things were really going wrong!

There are advantages and disadvantages to all the different types of evidence. Some types let you cover a lot of ground very quickly, but at the same time may come from a source that is either limited in perception, or perhaps even partial. To get the best from the raw data, a good rule of thumb is to train auditors to develop findings with at least two types of evidence to support it. Relying on one type alone, even documentary, will lead to an incomplete picture and could prove disastrous. Analysis of the evidence should rid the final report of any assumptions and such an analysis can only be carried out in a rigorous enough fashion when there is more than one type of evidence present.

Reporting: How can corrective and preventive actions be truly effective?
When analysing the evidence, it can be quite difficult to discover the exact meaning of the facts that have been found. Another useful analogy here is that of the auditor as a kind of vet. The auditor/vet examines a management system, diagnosing remedies for observable 'symptoms' and, just as with vets, their examination is not helped by the fact that the 'patient' can't tell them what is wrong. Just as with modern medicine, unless the diagnosis is accurate, it may suggest a remedy that appears to solve the perceived problem, while in fact merely suppressing it and leaving an incomplete cure. Under those circumstances, the symptoms may repeat themselves with an increasingly damaging effect on the system itself.

There is a useful mental model that will help in the diagnostic phase of the audit. When an EMS breaks down, it will allow something to happen in an uncontrolled manner, or in an unforeseen way. These are the 'symptoms' we've been talking about. These symptoms will occur in one of three areas, which can have a variety of names, but which we call 'Direction', 'Realization' or 'Delivery'. They provide a way of organizing the audit evidence so that corrective (or preventive) actions can be accurately targeted, and cure the root cause, not merely the surface symptom that the audit has uncovered.

Direction

For an EMS to work fully, it must have a clearly understood purpose. This relates directly to the identification and evaluation of the environmental aspects and impacts, as well as to the policy, objectives, targets and environmental management plan. Any problems in the 'Plan' part of the PDCA cycle will make themselves known as 'Direction' symptoms. In plain language, when an organization has a clear direction, it will know where it is going, what it is doing and who should be doing it by when. If it subscribes to a formal EMS standard, it will have all the elements (clauses) of the standard in place, and these elements will individually and collectively meet the requirements of the standard. To establish whether an organization has any problems in the area of Direction, ask yourself whether the symptom could be caused by any of the following:

- Are all the system elements in situ?
- Have the management clearly identified what it is they are seeking to manage?
- Have the management established their planning system elements properly?
- Do the elements of the EMS match the requirements of the standard (if appropriate)?

If the answer to any of these questions is 'No', then the evidence suggests that some kind of corrective action is needed in this area.

Realization

With the best will in the world, no amount of clarity of purpose will ensure that continual improvement is delivered. As well as knowledge of where the organization wants to go, the system must also provide enough operational controls to be able to manipulate the factors that have a direct bearing on the identified environmental impacts. A driver may wish to take a motor trip to a specific destination, but no amount of certainty about where one needs to go will make up for missing steering, brakes and motive power. Look at the evidence that the audit has found and ask:

- Is what needs to be done about this impact clearly understood and identified?
- Are all the implementation elements of the system installed?

CHECKING

- Are all the implementation elements working properly?
- Are there any missing procedures/controls?

If the answer to any of these questions is 'No', then the evidence suggests that some kind of corrective action is needed in this area.

Delivery

Knowing where you are going, how to get there and being able to pull all the right levers is certainly enough for many people. However, when levers are actually pulled, the acid test is to find out if it was effective in achieving the desired result. This is an area where assumptions are tested to the utmost. Well-thought through and carefully designed systems can still fail to deliver, because although there is control, the control itself is not effective under the particular circumstances the organization is in. The fact that something has worked in the past is no guarantee that it will work in the future. If there is an effective delivery problem, you should find it by looking at the audit findings and asking:

- Are the procedures and controls related to this impact being used?
- Are the procedures and controls achieving the desired result?
- Have the circumstances changed since the procedures and controls were devised?
- Have the procedures and controls themselves changed to reflect this?

If the answer to any of these questions is 'No', then the evidence points to some kind of corrective action in this area.

This mental model is something for auditors to use with their final non-conformity findings, when they have reviewed the evidence and are trying to refine the findings and any subsequent corrective or preventive action. It is not worth attempting to use the model before that point in the audit process as it may simply confuse, rather than illuminate.

One of the reasons why the auditors need to understand the precise nature of the EMS 'breakdown' is that they are in the best position to help the management in situ devise an appropriate response. As such, you may also want to devise your own criteria for what constitutes a 'major' non-conformity and train your auditors to identify them. Whether 'major' or not, if the finding develops into a corrective action, this means that the problem has already occurred and action is required to rectify the situation. As part of that action, there may be preventive measures in order to stop the problem recurring at a later date. Auditors need to appreciate that this is not the same as a discrete preventive action where the purpose is to stop an identified potential problem or hazard from realizing itself. The most effective audits result in the most tightly defined and clearly understood findings. It is only then that the subsequent remedial actions will become equally clear to the auditee.

The formality of the reporting that follows the audit will be mostly aligned with the needs of the organization. There is nothing wrong with an informal approach, though it should be noted that both ISO 14001 and EMS require that corrective and preventive actions be documented, mainly to facilitate follow up at a later date, which would include the top level management review of the EMS that we talk about in the next chapter. The minimum distribution for such a report is to the managers directly responsible for the actions required by the audit findings, and the person(s) responsible for the maintenance of the EMS itself.

What are the key skills to look for in an auditor?

This always seems such a vexed question until you turn the whole question on its head, and think about the type of person who would take the 'Worst Auditor in the World' Award. The best auditors are patient, thorough, analytical and have good communication and interpersonal skills. Some of these skills can be taught, but it is much better to start from a high standard. Independence and impartiality are factors that can only be supplied by the way the audits are administered. If you are pursuing formal standards, due to their requirements for periodic legal compliance evaluations, an in-depth understanding of relevant environmental legislation and its future development will be essential at the appropriate points in your audit teams. Again, it is worth using something like ISO 19011 as a starting point, but not as a rigid template.

What are the problems of integrating quality, health and safety and environmental audits?

Though this subject is dealt with in more detail in Appendix V, it is worth stating here that where integrated management systems cover health and safety, quality and environmental impacts, the biggest barrier to a successful audit is deciding on a clear scope that the audit team themselves understand. Having talked to some who have tried to carry out simultaneous audits of different strands of the same management system, the biggest problem seems to exist within the head of the auditors. Management systems overlap and so does evidence. An audit trail which started from an observation about quality can still lead to a problem with the EMS or even health and safety. This points to an even heavier emphasis on the analytical skills of the auditors, not in terms of gathering the evidence, but in terms of correctly identifying the nature of the findings themselves.

The deciding factors will be the confidence that the audit teams engender and the way in which the audit logistics are handled. Integrating several types of audit will put at risk not only the effectiveness of the feedback into one particular part of the overall management system, but as many parts of the system as are included in the integration. It is hard to see that many companies would expose the independence and effectiveness of their financial audits to such risks, and it is doubtful that many more would be willing to do so with other equally important areas of their organizations. Integration may bear fruit in the areas of logistics and resources, where there are obvious savings of time, effort and the avoidance of duplication. These savings will have to be balanced against the threats to the effectiveness of the audit(s).

What are the extra EMAS requirements?

EMAS is more prescriptive about the structure of the audits drawing many of its requirements from elements that bear a strong resemblance to those found in documents such as ISO 19011 (see above), a standard that can be regarded as 'guidance only' for ISO 14001 users. Other main differences are:

CHECKING

- EMAS has an entire annex on the subject of auditing (Annex II);
- EMAS puts a maximum figure on the audit cycle of three years (Annex II 2.9).

Narratives

Smallco

Smallco's EMS is now well established and needs its first audit. The Ops Manager, who has now attended an EMS Lead Auditor course as part of his own training, is acutely aware that those staff with auditing skills are very biased towards quality auditing and have poor understanding of the skills needed for environmental auditing. There is no budget to send the three staff who are designated quality auditors on a similar course.

Despite colleagues' advice that the easiest way would simply be for the Ops Manager to just get on with it, he is less than enthusiastic about taking on the whole task himself. This is not just for workload reasons. Plan B spreads the workload while extending the skills base of internal quality auditors. Working with the Quality Manager, the Ops Manager develops a hands-on practically focused internal training programme for the three quality auditors. Several exercises are written which extend their QMS skills and reveal some key differences between QMS and EMS. With the help of the major customer, who has nine internal quality/EMS auditors, most of the training work takes place outside the 'classroom' with a few short debriefing sessions to compare experiences and findings. Over a period of six weeks, the Ops Manager essentially builds a skilled internal team which he will manage to carry out Smallco's first internal EMS audit.

BIG Inc

The biggest problem confronting the project manager is setting the scope of the auditing programme for the company. The large number of home-based staff, the fact that many of the objectives rely on influence rather than direct control and the nature of the work that the company carry out all make this area difficult to establish. In the end, and after close consultation with the potential certifier, the project manager decides that home-based staff will be audited on a sampled basis, aiming to cover them using a mixture of site visits, questionnaires and random checks. The idea is to achieve 100 per cent coverage of home-based staff within the chosen audit cycle of three years, a period thought to be appropriate given the high number of indirect environmental impacts.

The evaluation of legal and other requirements is carried out at the office sites on an annual basis, given that there are no complex or toxic manufacturing processes taking place directly on the sites. However, the project manager also recommends that the board consider instituting a supply chain audit that does cover environmental legal compliance issues due to the regulations on the disposal of electronic waste. This audit may help the company to source materials that do not give their clients disposal problems at the end of the product life adding more value to their consultancy.

In the case of associates, the project manager decides to regard them in the same light as all other suppliers, and thus will not directly audit them. In recognition of their unusual position with regard to the company, associates are sent regular electronic questionnaires which require key details of their compliance with corporate environmental policy to be disclosed. In the subsequent assessment of the EMS by an external body, this scoping is found acceptable by the certification body, and the EMS is successfully certified.

Trade secrets

- In our experience, any documentation relating to contractors and suppliers is a particular auditing weak spot, as is anything to do with waste segregation and disposal.
- If your audit doesn't find any non-conformities, it doesn't mean you don't have any EMS problems, it just means you (or your external assessor) haven't found them, or they haven't happened yet.
- When giving findings to a management team, stay factual. If you don't think you can make it stick, make it an 'observation'.
- Objectives and targets can be changed to reflect business needs, but make sure that the system reflects the change in all the appropriate elements, including the environmental policy.
- You can't use an audit to get managers to support the EMS if they didn't support it when you were agreeing objectives with them.
- Good auditors strengthen the link between an effective audit and good company performance.

Things to think about

- The better the auditing process within an EMS, the more likely regulators are to be able to respond when it comes to calculating their own inspection regime. There has been a lot of speculation as to how and when organizations might be able to reap a 'regulatory premium' in terms of fewer external visits. Regulators themselves have been keen to explore this possibility. As yet, regulators are not convinced about the consistency of EMS implementation nor do they possess a mechanism that is sufficiently flexible to make such a proposition a serious possibility. It is worth keeping an eye on developments here as the overall situation may change on a national or even regional basis.

CHECKING

Acting

14

Management Review: Where Do We Want to Go Next?

Keeping an EMS alive and relevant to the organization's activities means more than just successful audits. Every now and again, it is good to take a step back and to take the long view on where you are and where you would like to go, not just operationally but strategically. It's a time to acknowledge your successes, think through some of the obstacles and renew the vision that brought you to this point in the first place. The most successful organizations have senior management who are always setting themselves new goals, as well as enjoying the achievements – and for this, environmental issues really do need room at the top.

✓ ISO/EMAS quick check

Area of EMS	ISO 14001	EMAS
Management review	4.6	Annex I-A.6
Evaluation of compliance	4.5.2	Annex I-A.5.2
Non-conformity, corrective action and preventive action	4.5.3	Annex I-A.5.3
		Annex I.B
	All other elements of an EMS (Indirectly)	All other elements of an EMS (Indirectly)

▢ Chapter executive summary

Try to answer all of the following questions in ways which make sense to you and your organization:

What is the purpose of a management review?
What are the inputs required?
What are the outputs?
How do you know when your management review is effective?
What are the extra requirements for EMAS?

🖎 **You are here**

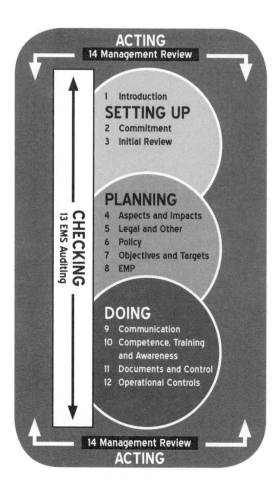

This is the only element that has been included under the 'Act' part of the PDCA cycle. Although it may look more like planning, once an EMS is established, this is where troubleshooting techniques mean effective action.

🔧 **Toolkit requirements**
To work through this chapter you will need the following knowledge and skills:

Knowledge

- strategic direction of the organization;
- changes since last review;
- auditing and compliance output;
- EMS performance since last review.

Skills

- ability to analyse/define business and environmental issues;
- ability to prioritize;
- conflict resolution.

 Context

In our experience, most people accept that change is inevitable. Even so, we also know that many managers might believe that when established, management systems will hardly need any maintenance. Looking around, it's hard to find something that follows this rule. If you want to turn your EMS into a managerial dinosaur, then don't have a management review or just go through the motions of one. There are plenty of other reasons, however, that make stasis an impossibility.

For those working with a formal EMS standard, the system is committed to the dynamic of continual improvement, and this alone will bring out the need to reassess on a regular basis. Those focusing on a 'legal compliance only' driver for their system will still need to make this assessment, mainly because implicit in their choice of pegging their environmental performance to obeying the law and not exceeding it, is the fact that the organization is committing itself to a constant, slightly worrying, game of 'legal catch-up'.

Obviously, it is not only the law that changes, but also the organization itself. New processes, new materials, new designs, new customers and new management initiatives all contribute to the pace of change within even the smallest companies. The average amount of time that a senior manager will now spend employed by an organization is three years and dropping; with each move, there is a slight change in the strategic direction and business priorities within the company. Changes at this level occur, if for no other reason, because different executives have different ideas, and everyone likes to make their own impression on an organization. At extremes, this constant stream of new ideas might lead to employees suffering from 'initiative fatigue'.

Even if your own organization is not suffering from this extreme form of problem, there will still be an element of flux in your operations, processes and overall circumstances which makes a management review of the system an imperative. The key word in the process is 'relevance'. Is the EMS itself relevant to what the company is doing? Procedures and responsibilities get dropped because they are no longer relevant to the immediate concerns, or no longer fit the circumstances that apply. The management review is your chance to take a bearing from a different map, and check that the relative fit between 'management' and 'managed' is still comfortable.

It is worth bearing in mind that many individuals feel as though they benefit directly from environmental management, unlike, say, quality management where the benefits are seen as accruing to the organization first and the individual second. This direct link will certainly promote active feedback through the auditing compliance process, but only the management function will be able to obtain the necessary detachment for the strategic view needed during this part of the management system cycle.

ACTING

Tasks

What is the purpose of a management review?

The purpose is to assess the continuing suitability of the EMS in delivering the organization's goals, as well as the relevance of the goals themselves. Think back to the role of the IER (see Chapter 3) and the way that it established what was actually happening, and what the significant environmental impacts were. In some senses, the management review is a more refined and ongoing version of that broad overview. For the environmental manager, or whichever team has that responsibility, this is the time to revisit the whole of that initial review. It is a good basis to start from, and will certainly ensure that the coverage of the interim review is at least as broad if not broader than the original. For more mature systems the results of previous internal audits and compliance evaluations will provide the data needed though these will need to be augmented by other inputs (see next section below). Using these types of templates will ensure that changes which may have escaped even detailed audits are an integral part of the management review.

When looking at the individual elements of the system, it is often worth revisiting the original risk assessment and impact evaluation work, to see if the procedures are still robust and applicable. All the other information made available may indicate that the system may well be functioning amazingly efficiently, but if the basis for calculating the objectives and targets has shifted and the environmental policy needs adjusting, then an efficient system will prove to be no replacement for an effective one.

What are the inputs required?

The inputs may be many and varied. As a result, they are best collected together in the form of a report which will help to organize the information in such a way as to make an overview possible. This is particularly important when trying to spot the interaction between two or more changing elements of the system or the organization. The contents of the report should allow the reader to carry out a gap analysis between the last management review, which with a new EMS is probably the IER, and the current one under consideration.

ISO 14001 and EMAS list a minimum series of inputs for the review, so we've rolled them up into our list below and added to them for added value and effectiveness. Think about the following elements:

Changes or planned changes to:

- the law and regulations;
- associated legal guidelines;
- 'other requirements' (corporate policies, trade association initiatives, codes of practice, etc);
- processes/materials/activities/site's infrastructure/plant, etc;
- supplier/contractor lists, or changes brought about by them;
- staffing levels, organization structure, personnel;
- the organization's identified environmental impacts.

Information from:

- audit reports (particularly corrective and preventive actions and their current status);
- reports on individual incidents;
- other information sources (trade conferences, press reports, professional journals, etc);
- market reports;
- environmental journals on emerging issues;
- internal environmental committees;
- external interested parties;
- complaint and incident records;
- planning meetings/updates of risk assessments;
- training records and needs analysis;
- previous EMS management reviews (if any);
- related reviews of quality or health and safety systems.

Assessments of:

- rate of continual improvement;
- extent to which objectives and targets have been met;
- performance of individual EMS elements;
- overall performance of EMS;
- continuing suitability of EMS.

What are the outputs?

These will be strongly linked to the management review report compiled in advance, which will not only provide a structure and agenda for the review meeting, but can even go so far as to structure the type of decisions that need to be made during the process. Again, ISO 14001/EMAS has a basic list of expected outputs that acts as a minimum requirement. As we think it is only sensible to go beyond the minimum for the extra effectiveness gained, we have included the stipulate elements in our list but not limited ourselves to them. Key areas which will be the subject of potential action, to maintain EMS effectiveness, as a result of the review can be split into seven headings:

- Policy: do we still meet our policy commitments and does our policy still meet our needs?
- Legal and other requirements: are we maintaining compliance and if required, could we use the outputs from our MR to demonstrate this to regulators?
- Evaluation of compliance: what is this saying about the organizational ability to maintain its performance within expected parameters?
- Objective and targets: which, if any, of our objectives and targets might we need to modify? And what new goals will we set ourselves?

ACTING

- Impact identification/evaluation: what are these data telling us about the effectiveness of our EMS?
- Incidents/system failures: what are these data telling us about the efficiency of our EMS?
- Monitoring failures: are monitoring systems adequate for our operations?

How do you know when your management review is effective?

There is not going to be one single indicator of effectiveness which will act as a litmus test in this area, no matter what type of system is being reviewed. If the EMS is still adequate in scope, effective in delivering against the targets drawn up by the organization and has targets that are still relevant, then the review will in itself have made its silent contribution. Be aware, though, that it is not the case that a 'good review' will mean no changes to the system. Simply because a review doesn't reveal the necessity of change, it doesn't follow that the system or the targets themselves cannot be improved. Much will have to do with the timing of the process.

Exactly when a management review takes place is very much at the discretion of the organization itself. Being the 'Act' in the well-known PDCA quality management cycle, the review comes at the end of a nominally complete cycle, indicating that the whole of the system can be refreshed by a strategic top-down perspective. EMAS puts a maximum figure on this cycle (every element in the system should have been audited – known as an 'audit cycle' – at least every three years). By comparison, ISO 14001 simply prescribes the need to demonstrate a planned review cycle. However, in practice, ISO 14001 certifiers usually advise their clients that as external assessors they will be using at least a two year audit cycle, because it fits in with the average ISO 9000 methodology. You could have a strategic review of the system once every month, if that fitted with your other management practices, but the depth of the review suggests that this would not be practicable. Every year would be quite sufficient for most organizations. Any earlier than that and the system won't have produced enough information and records to be worth looking at. Remember to document the conclusions of the review and, where necessary, communicate changes to the relevant staff. The measure of success beyond this point will be how many people understand the need for the changes that come out of the review, so clear communication internally and externally will be important.

What are the extra requirements for EMAS?

As we pointed out in the previous chapter on EMS auditing, the EMAS regulation is more prescriptive about the structure of an audit. In the same way, EMAS is more specific about auditing improved performance in the environmental area, and includes it in Annex II of the regulation. This will most likely have been addressed by anyone using ISO 14001 as a guide, but it is worth drawing the reader's attention to the requirements. EMAS is also more detailed concerning external communications, employees and the provision of information to the public.

This last point is a reference back to the environmental reporting elements of EMAS, the most significant difference between this scheme and ISO 14001. The reference in Annex II-2.7 ensures that the reporting does not simply stay at the level of issuing a statement on performance at pre-determined intervals, but encourages companies to engage in a continuing dialogue with external stakeholders. This is worth bearing in mind when devising objectives and targets for the company, and it may even be worth starting a formal committee or forum in order to keep such a dialogue going. Many managers may approach this with fear and trepidation, but if it is regarded as a way of engaging in meaningful discussion and not simply a way of parading prejudices, then more good may come out of it than one many think.

Cooperation with the public is also emphasized in EMAS, worded in a far more specific manner than the ISO 14001 requirements. Most responsible management will already have ensured that contact has been made, but instructions on the environmental aspects of the use and disposal of, say, a product may not have been included in the organization's environmental management plan, even if using ISO 14001. It may not have been considered a significant enough impact to warrant action until a later date in the view of the management, but for those involved in EMAS, the provision for it is written into the requirements from the beginning.

Finally, the wording concerning employee involvement means that, for EMAS, active steps have to be taken to ensure that they can participate fully. Simply showing the staff your environmental policy and asking them to indicate that they have read and understood it, may not be enough. The EMAS wording, however, certainly appears to put the onus of responsibility on the client company right from the beginning, so it is worth taking this into account.

 Narratives

Smallco

The EMS has now come full circle and it is time to carry out a management review. The hard work and energy of all staff, especially the Ops Manager, has paid off and paid back. There have been some excellent results in this first 18 months of operation including short-term cost savings, several of which have exceeded the targets set. As the management representative of the EMS, the Ops Manager has the responsibility for orchestrating the review meeting. An agenda, inspired by ISO 14004, *EMS, General Guidelines*, has been drawn up and circulated as has the EMS audit report with its straightforward two page executive summary.

One issue concerns the Ops Manager. It has surfaced that two managers feel that the EMS should be put on the back burner now that cost savings have been established. They are actively lobbying for this to happen. This is a critical time for the EMS and the Ops Manager has covered 'what happens next' in one of the appendices of the audit report. Always one step ahead, he has prepared a checklist of possible responses to the issue:

1   In this first year we have saved significant sums of money through simple measures.

2   In total, these savings mean we can undercut our competitors by 6 per cent on quotations.

3   We have negotiated a 12 per cent reduction in insurance costs because we have an EMS in place.

4   Our internal EMS audit has identified 37 areas for improvement which make business sense.

5   Five customers (*) have visited specifically to look at our EMS work and one has actively helped us through this first 18 months. We have letters on file from customers endorsing our good practice and have improved relationships and as a direct result of this work discussions are underway with one customer to develop new product related environmental objectives and targets. (* Three have their own certified EMS and require their suppliers to work in this area.)

6   We have been nominated for an environmental reporting award in our region. The media coverage of this produced seven enquiries about our work and one new customer.

7   The local regulator has signposted four businesses to our EMS and clearly regards us as an example of good practice. This is a massive improvement from two years ago when we were threatened with closure because of our poor environmental performance.

8   Productivity has improved by 4 per cent overall in the 18 months since we instigated an EMS. I am confident that the EMS, and especially our work to listen to staff ideas, is a significant contributory factor.

9   We now use 17 per cent less raw materials and 8 per cent less energy to produce the same product output. Our transport costs have fallen by over 5 per cent which includes fuel and maintenance costs.

10 We have reduced pollution in measurable ways throughout our operations.

11 In a survey of local school leavers, 54 per cent said they would prefer to work for a company which was serious about environmental management. Smallco was an unprompted example of good practice.

12 We can continue to improve our performance, both business and environmental, and show bottom line benefits in doing so.

Smallco have now completed Phase 5 of their phased implementation of EMS. Phase 6 – optional external assessment – is firmly on the agenda and the MD will communicate with all customers and suppliers to seek their views.

BIG Inc

BIG Inc has grown steadily over the last two decades, to the point where the company is now publicly quoted on the stock exchange (though 75 per cent of the shares are owned by the original management team). Such has been their recent success (profits have been growing by 30 per cent year on year over the past five years), that a Pacific-based international IT manufacturer has been showing an interest in a merger. The original team negotiate a staged 'buy-out' which will take their company on to the world stage, and the next phase of its development.

Such a move has an unexpected effect on BIG Inc's environmental policy and management system. Many expected the buyer's own environmental policy to supersede what had been developed for BIG Inc, but due to the successful nature of the work carried out by the project manager and their successful application for ISO 14001, the buyer sends over a review team from its own corporate department in order to learn what they can from the BIG Inc experience. The resulting report leads to changes in the larger corporation's policy, though there are some changes that have to be made by BIG Inc in terms of delivering specific policy commitments towards use of energy efficient hardware.

Trade secrets

- In larger organizations, a successful review can depend on your ability to 'manage upwards' by setting the agenda with a written report.
- Include an 'Executive Summary' at the front of the report and spend a lot of time condensing the issues so that they fit one side of A4 paper. This is often all a busy senior manager has time to read, so if you need to get something across, do it here. Technical information will still be needed, but include it as appendices.
- The most effective reports include a 'Decision Profile' for senior managers: a list of decisions that must be resolved by the end of the meeting.
- A gap analysis between current practice and the environmental review (or the interim equivalent) will highlight changes that need to be addressed.
- In smaller organizations, the need for formality is not so high, but don't forget to write a report and support documentation to indicate that at least the issues have been considered. The idea is to indicate that conscious decisions have been made, rather than the status quo having been preserved by default.

Things to think about

- Regulators are becoming increasingly keen to see evidence that organizations are dealing with legal compliance issues in an effective way. A thorough management review can easily provide such evidence and add depth to a simple evaluation of compliance. Such evidence may well result in less frequent regulator interactions with consequent savings of time and disruption and a less stressful time for the environmental management representative and their colleagues.
- The strategic element of the management review will become more important as organizations confront the rigours of sustainability and CSR. To a degree, it already combines economic and environmental concerns under the same roof, so it is a relatively small step to include social impacts. In the future, all three will have to be considered jointly without one taking primacy over the other two. Why wait when the mechanism is already there?

ACTING

Appendices

Appendix I
Understanding the Requirements of Environmental Management System Standards

Whether or not you are aiming to install an EMS that meets the requirements of the international standard ISO 14001, the EMAS, or a specific stage in a phased EMS implementation scheme (such as the UK's Acorn/BS 8555 scheme), it is useful to have a full understanding of how the basic elements of an EMS interact with one another. This appendix acts as a guide to the elements of the three types of scheme.

The international standard for EMSs was first published by the International Organization for Standardization (ISO) in 1996 and was last revised in 2004. Its official catalogue title is *ISO 14001: 2004, Environmental Management Systems – Requirements with guidance for use.* As part of a wider agreement, the standard has also been adopted as a European standard by the European standards making body, Comité Européen de Normalisation (CEN). Within Europe, it is therefore officially known as EN ISO 14001, to indicate its dual recognition. It is part of a series of standards produced by the ISO Technical Committee (TC) 207: 'Environmental Management'. The main committee acts as an umbrella to a series of subcommittees and working groups exploring the development of standards, technical documents and guides on the broad subject of environmental management. Standards such as these form the basis of industrial self-regulation, and are voluntary unless otherwise cited in legal instruments.

For anyone seeking to establish an EMS, it is worth knowing what other publications are available in the series. Covering supporting management tools as well as related management systems guidance, an environmental manager may well find them a treasure trove of reference material. Our opinion of the relevance of the individual standards to implementers is detailed in the last column of Table AI.1. Our intention here is to give an overview by listing the current subject areas for the benefit of those who wish to read further on a particular subject.

An important general rule is to read the Introduction and Scope sections of each standard, before getting into the detail of each one too deeply. They are easily overlooked, yet they will give the reader the context for the use of the standard; a particularly important point to bear in mind as all the standards except ISO 14001 in the series are 'guidelines', in other words, purely advisory when it comes to being assessed by an external body. Only

Table AI.1 ISO 14000 series overview

Standard number	Title	Relevance
ISO 14001	EMS: Requirements for use with guidance	Requirements are expressed in Clause 4. Useful additional guidance in annexes which are advisory only
ISO 14004	EMS: General Guidelines on Principles, Systems and Supporting Techniques	Useful background information on the approach to EMS installation
ISI 14015	Environmental Assessment of Sites and Organizations (EASO)	Guidance useful for self-assessment or pre-acquisition auditing
ISO 14020-14025	Environmental Labels and Declarations	Not directly relevant to EMS implementation. Useful only to those interested in taking part in an eco-labelling scheme, or in making declarations concerning products and services with environmental aspects
ISO 14031	Environmental Performance Evaluation - Guidelines for Environmental Management	Very useful in establishing measurements for objectives and targets and environmental performance indicators as part of an EMS, or as a precursor to the installation of an EMS
ISO 14040-14048	Life Cycle Assessment	Could be useful in getting to grips with LCA as part of your IER
ISO 14050	Environmental Management: Vocabulary	Advisory, but very useful to ensure that everyone is using a common terminology. Especially useful to those managing multi-site operations
ISO 14063	Environmental Communications	Guidance on the full range of environmental aspects of both internal and external communications

ISO 14001 is an actual specification, which means that 14001 is the only standard in the series against which an external registration or certification body can measure an organization's management system. Thus, if a company chooses not to follow the guidance in ISO 14004 to the letter, the certification body cannot respond that the company has in any way failed to meet the requirements of ISO 14001. Even within ISO 14001 itself, it is important to remember that the Annexes at the end of the standard are 'informative' (advisory) rather than 'normative' (non-optional).

Finally, though not officially part of the ISO 14000 series, a special working group of the technical committee did produce ISO Guide 64, entitled *Inclusion of Environmental Aspects in Product Standards*. Designed for the use of standards makers involved in writing national, European and International product standards, the guidance it contains is, perforce, rather broadly expressed.

Aside from ISO 14001 and its supportive series of standards, sites operating within the Member States of the European Union can choose an alternative voluntary scheme which achieves formal recognition not only of a defined environmental management system, but of a public report of that site's environmental performance. This is called the Eco Management and Audit Scheme (EMAS) and its requirements are outlined in a European Union Council Regulation (No. 761/2001). As a regulation, it is mandatory for each Member State to offer the scheme to certain defined strands of industrial activity. Participation in the scheme by the industrial organizations is, however, voluntary.

Whichever scheme is chosen by an organization, there is an alternative route to the same goal, namely making use of a phased implementation scheme. These are developed on a country-by-country basis, so check with your certifier or national accreditation body as to whether there is something available that might apply to your site(s). Such schemes provide a staged route that leads to :

- incorporation of formal EMS elements with recognition through external verification (inspection) at pre-determined points;
- ISO 14001 (if desired);
- EMAS (if desired).

This particularly suits Small and Medium Sized Enterprises (SMEs), larger organizations who may not wish to commit to complete EMS implementation at the start of the process or any organization that has limited resources available for the project. See below for further detail.

Do you know the structure of ISO 14001?

If you are going to be using ISO 14001 as a model for your own EMS (whether you use a phased implementation scheme or not), then it is worth familiarizing yourself with the overall 'shape' of the standard before you start. This will make it easier, not only when referencing particular requirements in the standard, but also when diagnosing where things might have gone wrong when non-conformities appear during audits.

Like most technical documents there is a certain amount of preamble to go through. When you open the document, the first element you will find will be the Foreword. It is tempting to ignore this, but there are some very good reasons to read it carefully as the Foreword may be national or regional, depending on where you bought your copy of the standard. It will help to put it into the context of other national standards and regulations. If you are operating several EMS across national (or even regional) boundaries, be aware that this context may change. After that, you'll find the following introductory elements before you get into the management system requirements of the standard proper:

APPENDICES

Introduction

This is a useful read, outlining the basic thinking behind the creation of the standard and introducing the reader to the model of continual improvement which drives the EMS structure. It also makes clear that the standard does not contain absolute requirements concerning environmental performance, other than the baseline position of the commitment to obey the law and prevent pollution which is part of the requirements in the policy clauses of the standard (see below).

Scope

Easy to skip over but worth remembering, as it establishes the scope of use for the standard, including the idea that organizations may wish to self-declare against the requirements of the standard as an alternative to seeking external validation and registration. If you are not sure what the standard can be used for, then check here.

Normative references

Though there are no other normative references at present, this will be the place to look for related standards references that have a direct bearing on the requirements of ISO 14001. As it is currently a 'stand-alone' document there is no other required reading.

Definitions

This is a series of definitions which sets the terminology for the purposes of the standard. It is worth checking here that the definitions are the same as those of your current company practice and culture, especially in the way the words 'objective', 'target' and 'procedure' are defined. If you find differences, care should be taken in the way that the terms are used, especially in EMS-related documentation. You may find the standard and your company divided by a common language.

Once you have read through these important and necessary opening elements, you will reach the central core of the document, Clause 4, *Environmental management system requirements*. The requirements are structured around five basic principles, which bear more than a passing resemblance to both the PDCA cycle of quality management, and the structure of the chapters in this book. The five principles are:

1   policy;
2   planning;
3   implementation and operation;
4   checking and corrective action;
5   review.

The clauses are numbered in sequence, with the addition of a special opening general requirements sub-clause, and as you can see, form the basis for an excellent maintenance plan for your own EMS once it is firmly established. Note that the changes to the PDCA structure in this book were deliberately undertaken to reflect the more open requirements of those who are starting down the EMS road, and have yet to install an EMS, or who may wish to do so using a phased implementation scheme or perhaps ultimately without using ISO 14001. The start-up sequence for an EMS is somewhat different from the maintenance cycle.

Informative annexes

These annexes do not form part of the assessable specification of the standard. There are three altogether, covering the following:

Annex A: Guidance on the use of this International Standard

The information here is structured to reflect the structure of the requirements themselves, so there is a direct correlation between the numbering system of both parts of the standard for easy reference. The clauses in Annex A reflect the numbering and structure of the main part of the EMS requirements: Clause 4 (eg Clause 4.6 and Clause A.6 both relate to operational controls). The guidance here is helpful, but is only of a very general nature. This book and ISO 14004 may prove more useful reference guides to have on hand, but remember that ISO 14004 is not a guide to ISO 14001, it is a guide to EMSs in general and the guidance should be taken in that light.

Annex B: Correspondence between ISO 14001 and ISO 9001

Set out in tabular form, the links are explored using both ISO 9001 and ISO 14001 as the baseline comparing the requirements of the standards. It is worth making two points: the first is that simply because the tables acknowledge that a particular subject area is covered in both standards, this does not imply that the requirements are exactly the same; the second is that there are many more detailed cross-connections between the two standards than could be shown in the table. The real extent and nature of the connections can only be established in the context of the individual organization and how they meet the requirements in the two standards. These tables are thus only guidance, not the final word.

Bibliography

This mentions specific standards from the ISO 9000 and ISO 14000 series which the TC felt might prove useful for users of the standard.

Do you know the purpose of each of the clauses of ISO 14001?

In official terms, the only body that is capable of 'interpreting' ISO 14001 is the technical committee that wrote it. In practice, every user of the standard will meet the requirements in their own way. To facilitate that, without going as far as 'interpretation', we've decided to run through each of the sub-clauses of the central Clause 4, and offer a commentary based on our own experience of what we know to be the purpose of the particular requirement in the context of the management system as a whole. The whole of this section will make more sense if it is read in conjunction with a copy of the standard itself.

Each user will have to look at the context of their own site or organization, see how far they are already going to meet the requirement and assess what needs to be done to close the gap. If you are going to consider registration by an external body, remember to check that you have the same definitions of certain crucial elements as they do (see Appendix II). You may wish to check your understanding of some areas with your appropriate national accreditation body, and your final choice of certification registrar will almost certainly be obliged to do so (see Appendix II).

ISO 14001 Clause 4 commentary

Clause 4.1 General requirements

These general requirements make it plain that in order to meet them, the proposed EMS should meet all the requirements outlined in Clause 4. It is also the first (but not the last) reference to a defined and documented scope for an individual EMS. Such a fundamental requirement ensures clarity from the very beginning as well as maintaining focus throughout subsequent implementation cycles.

Clause 4.2 Environmental policy

The policy is the public tip of the iceberg that is your EMS. Because it outlines your approach to environmental matters in a very general way, ensure that it reflects what is happening inside your system by keeping it up to date. If you do let the policy and the system drift apart, and you end up not doing what you say you do in your document, don't expect any certification body to hand you a certificate, or to let you keep the one you've got. If you haven't included specific objectives and targets in the policy, this will give the document a longer 'shelf-life'.

Ensure that the policy is defined by top management. It can be a little confusing as to who constitutes the most senior management, especially if applying the standard to a multi-site operation. As a rule of thumb, ensure that the top management as defined by the extent of the system has a hand in endorsing the environmental policy. Many registrars accept a Chief Executive 'signing off' the finished document.

There is nothing in the clause that mentions including the documented scope of the EMS in the policy document, just that the policy should work within the confines of that scope. However, if you think it adds clarity for your stakeholders (again especially on multi-site operations), there is no reason not to include it.

Don't forget the 'other requirements' in your commitment to comply with legal regulations and 'other requirements' to which the company subscribes. It's easy to leave out customer charters, local initiatives and the like, but ensure that they get a mention if they are relevant.

Providing a framework for your objectives and targets is a characteristic that may not be evident from simply reading the contents of the policy itself, and there is nothing wrong with this. What should be happening here is that when your objectives and targets are looked at, they should be addressing the issues you have raised in your policy. If you have more objectives and targets than policy, you're simply doing more than you say and most registrars will simply point out that this is not to your advantage. If you have less objectives and targets than issues raised in your policy, you aren't doing what you say you will and registrars are likely to regard that as a major non-conformity.

Policies not only need to be communicated to employees, but to those who work on behalf of the organization as well. This could mean associates, partners, sister companies or contractors and subcontractors, so make sure you use similar methods to get the message across to all the relevant people as well as those directly on your payroll.

Making your policy available to the public can be easy if you think about your stakeholders and who you have talked to in the past before writing your policy. Use a variety of methods to get that element of proactive distribution included here.

Clause 4.3 Planning

All the following sub-clauses under 'Planning' are a key part of the EMS installation process. The more time spent on getting the requirements in this section right, the more obvious the remaining requirements will be. It is tempting for many organizations to attempt to compress this section and get to the 'hands-on' management. There are very good reasons not to do this. In our experience, more major non-conformities happen in this section than in any other part of the standard, especially within first time EMSs.

Sub-clause 4.3.1 Environmental aspects. This is a key requirement, for the simple reason that organizations need to go through a process that identifies and prioritizes what exactly it is they are trying to manage with their EMS. If there is a flaw in this part of the installation and maintenance process, it won't matter how good the rest of the system is, it will be directing the management of the issues without having looked closely enough at what is being managed. Because no two organizations are alike, the aspects they will manage will not be alike, let alone the methods they employ to undertake that management.

In this way, although ISO 14001 is not prescriptive in terms of laying out issues to be managed, it certainly directs organizations to become more aware of the environmental aspects of their operations. As this aspect of business has previously been left unacknowledged, it is easy to see why organizations have such difficulty adding this dimension to their business decisions. The requirement is also often seen as one of the biggest challenges for any company without a record or experience of environmental risk assessment. However, it is easier to start with very simple models of risk assessment and then allow them to get more complex over a period of time, rather than go for intense complexity from the very beginning (always depending, of course, that your operations and activities are relatively straightforward as well).

It is always a good idea to carry out a risk assessment as part of the initial environmental review, the baseline assessment that will support the development of this procedure. It is also a good idea to update that work as part of the regular management review (see Clause 4.6 below), as this will help to ensure that you are meeting the requirements of Clause 4.4.7, *Emergency preparedness and response* (see below).

This clause also pushes the boundaries of the EMS by stressing the need to consider not just those areas that you can control on site but also those areas that you can influence. Examples might be the environmental requirements you have placed on suppliers and effective eco-design can reduce the environmental impacts of a product when it is in use by a customer.

APPENDICES

One of the most difficult requirements to meet over time is the need to keep this process up to date. Possibly because many senior managers delegate the responsibility for developing this procedure to more specialized functions, they lose touch with the fact that the process has to be incorporated into many of the strategic decisions taken by the company from then on. All too often, organizations set up an EMS and then start to carry on with their business development process as though it is totally separate from the consideration of the environmental impacts. Ensure that at least one senior manager understands the process and that business planning and development processes are changed to reflect the ISO 14001 requirements in the future.

Sub-clause 4.3.2 Legal and other requirements. You may well want to keep a record of some kind in terms of acknowledgement of the relevant legislation. Your policy will have committed you to obey the relevant regulations, and this record will help to protect you from the unreasonable demands of partial interest groups. However, a written record is not the only way of meeting this requirement, which asks only that a procedure for identification of legislation be established and maintained. Again, thinking of the purpose of this requirement, how to meet it should become quite clear. Though ISO 14001 makes no prescription concerning environmental performance, it does provide for a reasonable baseline with this requirement in conjunction with sub-clause 4.5.2, *Evaluation of compliance* (see below). It would be hard to define an EMS as adequate if it allowed repeated breaches of the regulations. The requirement removes from the organization the defence of 'ignorance' which would probably not have been recognized in a court of law in any case.

For companies and organizations with a pre-existing background in 'compliance auditing' this should cause few problems. However, they should be aware that environmental law is slightly different from many other areas of the law in that it is linked to international agreements much more closely, is much faster moving and is still open to wide interpretation by many regulators. For those without such experience, focus not only on how you know which law is relevant but also on how those operating all the relevant procedures know as well. Think too about how you will keep the information up to date, extremely important in such a fast moving and fluid area. Don't forget 'other requirements' where they apply to your organization (see Clause 4.2 above).

Sub-clause 4.3.3 Objectives, targets and programmes. This clause is tied directly to the previous two sub-clauses when it comes to meeting the requirements. However, it also allows for other much wider considerations to be taken into account. While the previous two clauses are obliged to be free of the taint of business considerations (they are to do with accurate identification rather than a balance of probabilities), here is where financial, technological and organizational requirements are blended into the decision process.

The key at this point is 'prioritization'. There are many different parties involved even within the organization, where each function has its own priorities and will not be shy about presenting them. The financial department will have a close eye on budgetary restraint, the sales department may wish to improve the product first and foremost, while production operations may simply be focused on the final amount of product throughput.

Meanwhile, outside the organization, each pressure group or interested party may well have its own agenda to pursue. All this has to be balanced in a series of objectives and targets that are meaningful to the circumstances in which the organization finds itself. Hence the lack of prescription for ISO 14001 here, mainly because no two organizations are going to be in the same position at any one time. An individual response is required.

Remember that no objective and target can allow the law to be broken, pollution to happen, or be in direct contravention of what is stated in your environmental policy, including the commitment to continual improvement. Only when you are sure of this can you turn your attention to the action programmes that deliver them.

The biggest single problem with EMPs is getting them to coordinate with existing work in such a way that all those responsible can acknowledge what their role is from the very start. For many managers and staff, environmental responsibilities are new and different; it is a human tendency to default to what we know rather than what is new, so expect a few problems at first and audit this area carefully while the system remains immature (for about the first 18 months).

Timeframes will need to be negotiated with those on the ground, who ultimately have to do the work. When timeframes change owing to some other business circumstance (ie objectives get deferred until a later date) there is nothing wrong unless the rest of the system fails to reflect this change. Also ensure that timeframe changes don't interfere with legal liabilities.

As the system 'beds in' at the operational level, a new problem with the programmes will become obvious: how is the programme updated to take account of changing operational conditions? A change in a material or supplier of sub-components can have a large effect on the programme without it being immediately obvious. Ensuring that new developments in such a wide area of operations get flagged up to the EMR can sometimes be difficult. Persevere, however, or the EMS will become less relevant over time to the business it is supposed to be serving.

Clause 4.4 Implementation and operation

No one should be surprised to find that there is the greatest number of sub-clauses in 'Implementation and operation'. This is where the day-to-day operations of an organization make direct contact with the system, and it ceases to be a 'paper only' entity. As a result, when installing a new EMS, review the existing methods and procedures carefully. By taking the line of least resistance, and adding to what is already there, you'll keep additional procedures and methods to a minimum. This not only minimizes your own work as the installer of the system, but also, more importantly, minimizes the risk factor of staff forgetting new procedures.

Sub-clause 4.4.1 Resources, roles, responsibility and authority. By the terms of this sub-clause, it is possible (and may even be desirable) to have more than one EMR, thus splitting the responsibilities for the maintenance and upkeep of the EMS. This should only be applied with care. Under certain circumstances, where the nature of the operations or the site indicate a diffuse management structure, it may be beneficial to have such a split. Unfortunately, in our experience, such a division of responsibilities is not always carried out cleanly, leading to continual confusion as to who should be doing what. The potential risk of a breakdown in the system is therefore much higher.

Whatever the final decision here, organizations are required to 'ensure the availability' or adequate resources in order to keep the EMS effective. This means that there should be some place in the system that defines what contingency plans are in place to ensure the continued supply of resource (ie what happens when the environmental manager is off site, for example).

It is pretty obvious what would happen without meeting the requirements of this clause. If any of the clause title items are ill defined or unacknowledged, the result will be a potential failure to deliver the desired results. Stating the obvious, your response to the requirements should be designed to avoid this.

This is also another clause that specifically mentions top management; this time with a view to involving them in the selection of the EMR. This does not, of course, debar one of the directors or senior executives themselves being appointed the EMR, and there may be strong reasons to do so in a smaller organization. After all, directors bear many of the liabilities when operations go wrong, so it may be in high risk operations they will gain more comfort being more directly involved than simple delegation.

Sub-clause 4.4.2 Competence, training and awareness. It is worth noting that this sub-clause begins with competence and stresses the type of training that is needed, as much as the fact that the organization has to identify its specific training needs with relation to environmental impacts. In an EMS, training is the ultimate preventive action, and in that sense, there is much to be gained from tackling your training delivery by prioritizing it according to the level of risk from procedural failure. Don't forget that everybody will require some form of 'awareness' training, and that where environmental impact training is received, it will need to inform the recipients of what will happen when procedures aren't followed, as well as when they are. Even organizations with low staff turnover will need to ensure that new employees are included in the TNA as well as the catch-all induction training sessions.

Another element of this clause that bears careful scrutiny is the reference to 'any person(s) performing tasks for it or on its behalf'. This pushes the requirements beyond the boundary of employees into areas that might easily get overlooked; not just contractors and subcontractors, but remote workers, temporary staff, associates and partners.

Sub-clause 4.4.3 Communication. The largest potential problem in this area is that a simple breakdown of internal communication will lead to a bigger breakdown within the EMS itself. You can guard against that by thinking of the significant impacts of the organization and aligning communications needs around these in the first instance. New developments are the next area of focus, though this will be more likely to be at a strategic level. Though you may not want to stifle independent action or the need for individuals to respond with flexibility to changing circumstances, neither do you want them to imagine they can act in a vacuum without any consequences.

External communication will be aligned with the culture of the organization. Even if it isn't already, this sub-clause brings the decision making process about the subject into the area of the conscious, rather than simply leaving it to a default position due to an oversight. If the decision to communicate is a positive one, then there should be a follow through in the shape of evidence of the methods the organization has or will use.

Sub-clause 4.4.4 Documentation. The requirements here seem at first sight to be unambiguous. However, we have always found areas that present the EMS installer with exercises in interpretative judgement. Most of the documentation is defined by the scope of the EMS itself (which, of course, will be on a document). However, the interrelationships between documents that are internally derived and those that emanate from external agencies often do not match up. The interface between these two areas is worth examining in detail.

Similarly, ensure that your ideas of what constitute the 'main elements' of the EMS match with your external assessor. Again, if you can extend existing documentation in your quality or health and safety management system, don't be afraid to do so. Make sure that if you do, the scope of that documentation covers the areas that you need it to, especially as an EMS can affect a much wider range of actions and activities than, say, health and safety guidelines, which may only focus on high risk areas.

We have always regarded such documentation as a way for new employees to find their way around the system quickly and easily should they need to, or for existing employees to find out what they need to know when operating outside their own normal area of operations. You can always 'road test' your final documentation on someone who doesn't know the organization or the site very well, and see what feedback they give you.

Sub-clause 4.4.5 Control of documents. Documentation in an EMS can be much more than a bureaucratic nicety; with such an emphasis on the law in environmental protection, documents can be the ultimate in terms of protection. This puts an even bigger emphasis on document control. Those with QMSs may well have a head start in this area over those who don't, especially as the organization is required by this clause to have a procedure that approves documents 'prior to issue'. Even so, when adapting QMS document control procedures, legal documents or any legally related material such as calibration and monitoring records should be treated with special attention and care, especially when it comes to the removal of obsolete material and the archiving of records. Remember that external documents will need controlling too.

Sub-clause 4.4.6 Operational control. This sub-clause covers one of the most crucial links in the chain, so it pays not to overlook the obvious. We really have come across EMSs that have wonderfully defined targets for their staff, with a well-conceived and thought through management programme to deliver it all. Staff have been appropriately trained, procedures written and responsibilities defined. The only thing that appears to have been forgotten is that the actual levers of control are not present, or that those given the targets don't have (or are not allowed to have) their hands on the levers.

Usually, when there is a non-conformity against this clause, the organization hasn't done anything that is very wrong; it just hasn't done enough that is right. This can be particularly evident when changes in operations further back up the production line have not filtered through to those given related targets. Our advice is to double check the obvious

APPENDICES

and make no assumptions here. However, we also know that we are all blind to our own assumptions in advance – they look like facts to us until proven otherwise. This is why impartial and independent auditing can always add value.

Sub-clause 4.4.7 Emergency preparedness and response. Don't be fooled by the position of this sub-clause towards the back of the standard; it's an item you need to tackle first, at about the time of the IER (see Chapter 3) when you are considering the potentiality of some of your major impacts. Health and safety studies may well help to highlight areas of risk, but remember that the focus of the environmental version is control of impacts to the environment, not to the risks presented to people. The clue, however, is in the title; it includes both preparedness and response. Interestingly, even when some organizations have carried out environmental risk assessments, their follow up to such a review has been to concentrate on either prevention or mitigation, but rarely both. It pays to think not only what happens when things go wrong, but what happens when your actions to put them right go wrong as well.

Clause 4.5 Checking

The title here is slightly misleading, in that 'checking' implies an activity that takes place 'after the fact'. It's worth contemplating that preventive actions are also woven into the requirements of the sub-clauses throughout this section. In environmental terms, waiting until a corrective action is necessary could be too late to prevent a pollution incident impacting on the local or regional environment, hence the need to spot potential problems.

Sub-clause 4.5.1 Monitoring and measurement. No surprises here, providing a little thought is applied in the way that you conduct your IER. Once you have identified your significant impacts, how do you know which characteristics need monitoring and measuring to ensure that legal requirements and other policy pledges are met? One of the most common problems that we have seen in new EMSs is that the monitoring regime isn't up to the level that the environmental programme would expect. In other words, objectives to lower certain heavy metal contents of water discharges are only going to be traceable if your monitoring equipment can detect the materials at the required levels of sensitivity, if they are calibrated regularly, positioned properly and checked at appropriate intervals, and the results recorded. Miss out any part of this process and all the work done further back in the production process is in effect uncontrolled, simply because you cannot verify the results.

Sub-clause 4.5.2 Evaluation of compliance. No one would question the wisdom of carrying out a periodic evaluation of compliance, either with the law as the focus or with other environmental requirements to which the organization subscribes. However, whereas many EMS installers may come across advice to incorporate this evaluation into their internal audit cycle, should they follow the advice, they may fall foul of their own certifiers/registrars, as well as suffering ongoing problems. During an initial assessment, external assessors will usually understand if the internal audit cycle has not been finished. On the other hand, they will be unlikely to issue a certificate if the compliance evaluation exercise has not been completed. Even when the certificate has been gained, there are distinct advantages to keeping the auditing function separate from the periodic evaluation (see Chapter 13 for more details).

Sub-clause 4.5.3 Non-conformity, corrective action and preventive action. Anyone used to the PDCA cycle associated with QMSs is going to feel right at home with this sub-clause, though it is worth remembering that there is a heavier emphasis on the preventive aspect of the checking that is required. This can mean paying closer attention to risk assessments during change management exercises as well as incidents and departures from procedures even if they didn't ultimately result in an environmental impact or a pollution release. Robust reporting procedures concerning these incidents are a must, simply because a system audit programme, no matter how well devised, cannot cover every part of the system at once.

For those unsure of the difference between corrective and preventive action, they can be loosely defined as follows: the first occurs after a non-conformity has occurred and further repercussions require action, while the second is an action taken in order to prevent a potential non-conformity from happening in the first place. In the case of either action, remember to keep your documentation up to date as the system changes to make accommodations of this sort, it's all too easy for a living system to evolve on the ground, without reflecting the changes in the documentation. This produces more potential problems in identifying non-conformities at a later date.

Using ISO 19011 as a template for your own developing audit programme can be extremely useful, especially in the early days of implementation.

Sub-clause 4.5.4 Control of records. Again, anyone versed in QMSs will recognize these requirements, and the procedures used within a formally recognized (ISO 9000) QMS will undoubtedly form the backbone of the EMS record keeping procedures, with one very important exception. Because of the close link with the law mentioned previously, there may be some legal requirements on the amount of time that records are retained. Ensure that your regime takes account of this even though record retention times are not an ISO 14001 requirement. Records in this case, doesn't just mean monitoring records, or environmental performance records, but also things like auditing records, training records and anything else that is used to support the EMS, which can be used to track the system performance over the past months and years.

Sub-clause 4.5.5 Internal audit. Though this is a relatively short sub-clause, it refers to a vital part of the EMS as a whole, and manages to pack in quite a lot of requirements into a small space. If you have used ISO 9000 QMSs auditing techniques, you will certainly not find anything unusual about the requirements, but be aware that as the scope of your EMS is, in all probability, different from your QMS, it will require a more extensive auditing programme in order to cover all the bases.

Remember too that the aims of the audit as expressed are that the EMS itself conforms to the requirements of the standard, the requirements of the company itself, and

APPENDICES

that it is being properly implemented (do you do what you say you do?) and maintained (have circumstances altered requirements? If so how and does the EMS reflect this?). The information will also go towards the management review of the EMS as well as the functional management of the company in order to carry out corrective and preventive actions.

You will get the best results from an informed but impartial and independent perspective, so remember to make sure that you can demonstrate that these characteristics are present in your auditors.

Clause 4.6 Management review

Many EMSs have been externally assessed without this review having taken place, though it would be very unusual for a certificate to be issued without at least some evidence that one is at least planned within an appropriate time-scale. If it can be incorporated into other review processes, so much the better and many of these will be on an annual basis, whether the cycles are calendar or financial. Again, note the words 'top management' here and the fact that the review itself should be documented, though many people rely on the formal minutes of a review meeting to provide this record. As the decisions reached in the review also have to be documented, such minutes will be examined by external assessors, so it is worth ensuring that there is some form of easily traceable link between the review documents and any subsequent changes to the system.

Remember that even though the inputs and outputs of the review are listed by the standard in detail, this does not limit the management team who may want to consider a larger number of issues. The review can be a chance for top management to assess the continuing strategic value of the environmental policy and the EMS that supports it. Alternatively, it can be a hollow and meaningless exercise carried out solely to meet the requirements of the standard.

For a quick reference to the requirements of ISO 14001: 2004, Box AI.1 has been provided for you to photocopy and keep with you. Originally devised by Alan Walch of the British Standards Institution's certification service, it's a handy reference for those who are just beginning to get to grips with the structure of the standard.

What do you need to know about the European Union EMAS Regulation?

The current EMAS regulation does have important differences from the requirements of ISO 14001 of which environmental managers need to be aware. The most important broad differences to bear in mind are that:

- EMAS requires a verified public statement of the site's environmental performance (usually annually unless you are an SME);
- EMAS specifically requires that an IER takes place;
- EMAS is more prescriptive in terms of issues covered by the IER and the structure of auditing process.

It is worth bearing in mind that the structure and the language of a piece of European law, which is what EMAS is in effect, is quite different from that of a standard. For anyone

Box AI.1 ISO 14001 quick reference

General requirements
4.1   •   **documented** EMS scope

Environmental policy
4.2   •   **appropriate** to environmental impacts of activities, services, products
    •   continuous **improvement** commitment
    •   commit to **legal** requirements
    •   set **targets and objectives** – review these
    •   **document and communicate** to employees, and available to public

Planning
4.3.1   •   **procedures** to identify **environmental aspects** of activities/products/services
    •   identify aspects which can have significant **impact** on environment
4.3.2   •   procedures to identify and access **legal** requirements
4.3.3   •   establish **targets and objectives** consistent with policy
    •   consider views of interested parties
4.3.4   •   establish an **environmental management programme** designating responsibilities and the means

Implementation and operation
4.4.1   •   define and document **structure and responsibility** for effective environmental management
    •   provide **resources**, skills, finance for EMS
    •   appoint **management representative** to ensure implementation and report to management
4.4.2   •   **training, awareness and competence** of personnel with significant impact on environment:
    (a) importance of **conformance**
    (b) possible **adverse impacts** of work and benefits of improvements in performance
    (c) their **roles and responsibilities**, including emergency preparedness
    (d) potential **consequences** of departure from procedures
4.4.3   •   establish procedures for **communication**, with internal and external interested parties
4.4.4   •   **EMS documentation** – be maintained and point to related documents
4.4.5   •   **document control** – available/reviewed/revised, remove or identify obsolete documents
    •   documents to be **dated** with dated revisions
4.4.6   •   ensure **operational control** over activities with significant environmental aspects
    •   activities, including **maintenance**, should be planned to ensure specified conditions met
    •   establish procedures where required to **prevent deviations** from EMS policy
    •   define operating criteria and include aspects related to **suppliers and contractors**
4.4.7   •   establish **emergency preparedness and response** procedures
    •   **test**, review and revise such procedures

Checking and corrective action
4.5.1   •   **monitor and measure** EMS activities, record and **track performance** and compliance
    •   monitoring equipment shall be **calibrated** and maintained
4.5.2   •   nominated staff investigate **non-conformances** and take action to mitigate any effects
4.5.3   •   **corrective/preventive actions** – appropriate to magnitude of impact
4.5.4   •   EMS **records** to be maintained and readily retrievable, defined retention periods
4.5.5   •   conduct periodic **EMS audits** to determine that planned arrangements, and the requirements of the Standard are being met
    •   audit programme to be **comprehensive** and based on environmental importance

Management review
4.6   •   **review effectiveness** of EMS, consider changes, commitment to **continual improvement**

APPENDICES

coming to the document for the first time, most of the useful information of the regulation lies in the annexes that are issued with the regulation, while the body of the document itself is expressed as a legal requirement upon the governments of the Member States.

As the regulation has evolved, it has been revised in such a way that ISO 14001 provides the structure for the EMS elements of the scheme. The current structure is shown below:

Article 1: The scheme and its objectives
Article 2: Definitions
Article 3: Participation
Article 4: Accreditation system
Article 5: Competent bodies
Article 6: Registration of organizations
Article 7: List of registered organizations and verifiers
Article 8: Logo
Article 9: Relationship with Standards
Article 10: Relationship with other EC environmental legislation
Article 11: Promotion of participation (especially SMEs)
Article 12: Information
Article 13: Infringements
Article 14: Committee
Article 15: Revision
Article 16: Costs and fees
Article 17: Repeal of previous EMAS regulation (No: 1836/93)
Article 18: Entry into force
Annex I: Environmental management system
Annex II: Environmental auditing
Annex III: Environmental statement
Annex IV: Logo
Annex V: Verifiers (competence, etc)
Annex VI: Environmental aspects
Annex VII: Environmental review
Annex VIII: Registration

This structure and its requirements make it possible for a company using an ISO 14001 system registered to the standard by an external certification body to have the system taken into account during the verification process for EMAS.

In general, EMAS allows companies and organizations flexibility in the way they meet the requirements. This applies especially to the structure of the public statement of environmental performance, the scope of the initial environmental review and to the scope of the auditing procedures. To access the regulation and a corrigendum published on 4 December 2002, readers may wish to visit the official web site to download their own documents. It can be found at http://europa.eu.int/comm/environment/emas.

The regulation applies to all organizations, regardless of industry sector. In addition, though originally site specific, it is currently applicable across entire organizations. However, for obvious reasons a single registration held by an EMAS Competent Body (national or regional administration) cannot exceed the national boundaries of a European Member State. Large organizations will need to take note.

SMEs, on the other hand, will definitely be able to use the increased accessibility and flexibility of the regulation. Fewer obstacles to involvement may not be the same as positive promotion and reduced fees, but each Member State is required to tell the Commission what it is doing to encourage take-up of EMAS. What may look daunting at first may well turn out to be less of a problem for smaller organizations than for larger ones.

Either way, organizations may find it easier to use the 'stepping stone' approach and acquire ISO 14001 before going on to EMAS or perhaps utilize a phased EMS implementation scheme and proceed to EMAS via that route. New disciplines mean new skills, and timing the application of resources in a development such as a new EMS could well be the deciding factor as to which route and which ultimate destination is chosen.

What are EMS 'phased implementation schemes'?

Phased EMS implementation schemes are precisely what the generic term implies: a complementary route to the installation and maintenance of EMSs. They do not seek to replace ISO 14001 or EMAS, but to clear away some of the obstacles that some organizations have found in the implementation process. They provide an alternative route to achieving the same goal, namely effective and appropriate environmental management of organizational impacts.

Most of them use the British Standard (BS 8555), or a similarly structured document as their basis and as these are usually issued in the form of guidance, the scheme usually incorporates a separate but related external inspection process. As the title of the British Standard implies (BS 8555: 2003 *Environmental management systems − Guide to the phased implementation of an environmental management system including the use of environmental performance evaluation*), the implementation process is broken down into a series of different phases, and at pre-determined points in the process, external inspectors can therefore verify that the management elements required exist on the site.

Each phase is broken down into a series of stage profiles, at the end of which there are stated achievement criteria. It is against these criteria that accredited inspectors, frequently the same personnel who are already employed by certifiers or registrars, measure the organization. The inspections can then build towards ISO 14001 certification or EMAS verification, if desired, but the great benefit of these schemes is that SMEs or organizations with simpler processes and activities on site, can choose to maintain their management approach at a more appropriate level (ie Phase 3) indefinitely. Annual visits by inspectors ensure that the organization is maintaining the system at the appropriate level in relation to their impacts.

Progress through the phases can be at a rate determined by the organization itself. Such progress is obviously influenced by resource availability, customer requirements and other stakeholder expectations. The key word in these schemes, however, is 'flexibility' and though originally designed with smaller organizations in mind, many larger corporate structures have made use of its approach.

The overall approach in the early stages is to involve the organization in management of their environmental impacts by using basic EPE techniques. This means that the commitment to improve is gained and can be acted upon straight away, without waiting for all the EMS elements to be in place. It also means that organizations have access to the business and environmental benefits at the earliest possible stage. This aspect is something that any EMS implementation project manager will appreciate, as it can be very difficult to keep environmental issues to the forefront of the business agenda while waiting for an EMS to produce results. 'Low hanging fruit' sounds like a disparaging term, but can be very important in gaining and maintaining momentum in terms of management system development.

This 'front loaded' approach to EMS implementation means also that, as the organization continues to progress through the phases, there is an increasing amount of formality and internal assurance that builds on the initial successes. With the initial emphasis on environmental performance, you may discover that there are requirements built into the schemes that are over and above what might strictly be required by ISO 14001. Examine the differences closely, but in the best designed of the schemes, these requirements will only be there to facilitate the flexible implementation structure and be based on best environmental practice. They may be extra, but they should still make good sense.

The details of individual schemes do vary, however, so ensure that you understand the approach and the limits of the particular scheme that you are investigating. You may find that major customers are willing to support such schemes with resources through a supply chain initiative (especially those who are registered to ISO 14001 or EMAS themselves) and the inspection regime associated with the schemes can also be used to check compliance with customer environmental contractual requirements. You may also discover that there are local, regional or national funding packages available.

One of the most appealing aspects of such schemes for hard-pressed management is that the initial level of commitment required extends no further than the first phase of the scheme, if desired. Further commitments to progress through the scheme can then be made depending on a variety of business factors such as market pressure, the availability of resources and perhaps the increasing complexity of operations or the applicable regulatory framework. If this is attractive, double check the structure of the scheme for anything such as imposed time-scales for progress – they may not exist in the scheme, but they may be imposed by funders or even customers.

Either way, phased EMS implementation schemes provide a useful way of introducing and embedding an EMS over a (potentially) extended period of time. You will still find all the guidance in this volume useful, though you may access it in a slightly different order to the chapter structure that we have chosen.

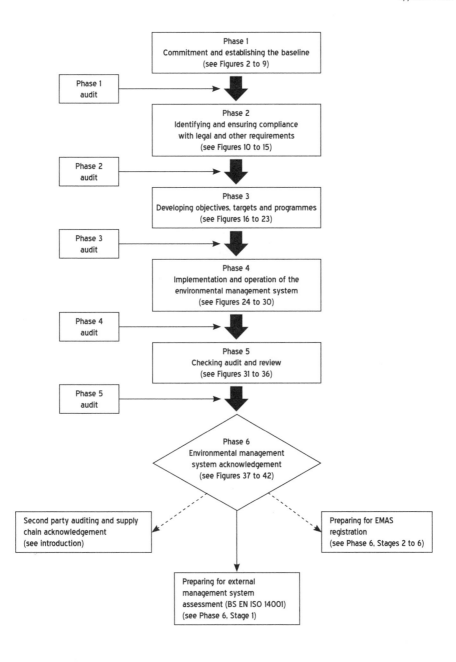

Source: Permission to reproduce extracts of BS 8555: 2003 is granted by BSI. British Standards can be obtained from BSI Customer Services, 389 Chiswick High Road, London W4 4AL. Tel: +44 (0)20 8996 9001. Email: cservices@bsi-global.com

Figure AI.1 Figure 1 from British Standard BS 8555

APPENDICES

Appendix II
Third Party Auditing

This appendix is not intended to be a comprehensive guide to third party assessment or 'external auditing' as it is sometimes called. Rather it has been included as a series of pointers for those EMS implementers who may wish to go forward to have their EMS assessed by an external third party. The reason to do this is to have the system (or intermediate system elements if they are considering a phased implementation scheme) formally inspected, certificated and registered. The benefits of such certification are separate from the benefits of installing and running an EMS within an organization, and should be considered separately by the top management of the organization. Even though many an EMS implementation project may have its success criteria defined as 'get us the certificate', top management should be able to look deeper than such a simplistic definition and make reasoned arguments for or against the certification process. Unless otherwise indicated, throughout this appendix, the word audit can be taken to include those inspections undertaken as part of a phased implementation scheme as well.

In some ways, external audits tend to be taken too seriously by those who are the subject of them, especially when personal performance targets and deadlines for registration to formal standards are concerned. We'll let you into a secret: there is no way to fail an external assessment. Assessment for registration is just that: assessment. The results are not expressed on a 'pass/fail' continuum, but as a series of diagnostic findings. It is true that if the assessors discover many non-conformities, or non-conformities that are major in nature, you may have a lot of work to do to get things back into conformity, prior to which the certificate cannot be issued; but if the aim is to have an efficient and effective EMS, isn't that what you wanted in the first place? Even with inspections under a phased implementation scheme, you will in all probability be given the chance to rectify the situation without having to submit the organization to another visit.

At worst, you will get valuable information on the state of your management system from the external assessors. This will take the form of an agreed set of corrective actions that will primarily fine tune your system to a pitch of effectiveness that will deliver the benefits you have set your sights on. As a secondary function, when the specific points raised have been addressed, you know that the assessors will be satisfied to the point where they can recommend the organization for certification. Even if an entire system is in such a state that it is not worth assessing until certain fundamental work has been carried out, most assessors will indicate this to the organization and allow you to call a halt to the assessment process until the work is complete.

What are the benefits of inspection certification/registration/verification?

If you are already benefiting from the installation of an EMS, if an energy management programme, a waste minimization programme and certain process changes are already netting you solid business benefits, the question about the benefits of external certification is more difficult to answer, and will depend very much on your individual circumstances.

The most obvious benefit is that certification gives an important ability to demonstrate that you do what you say you do, in terms of your EMS. Unless you are taking part in EMAS, or in some phased implementation schemes, this demonstration won't extend as far as actually auditing your company's environmental performance, but it will signify that your EMS conforms to the stated standard. This in turn may illustrate to your markets, whether at home or abroad, your financial advisers, your insurance companies and perhaps even your regulators, that you are serious about the management of your environmental impacts and your exposure to environmental liabilities.

The finance sector already takes certification to ISO 14001 and/or EMAS into consideration when it comes to placing companies on approved investor lists, or extending loans for individual projects that have environmental liabilities and risks associated with them. Regulators, although not going so far as to extend some kind of 'regulatory holiday' towards individual registered companies, have at least indicated that a certified EMS is certainly something that they too take account of during inspections, even if it is not formal recognition. Insurance companies too include certification to ISO 14001 in their criteria of risk assessment when creating individual policies, though again, there is no established direct link between certification and a lower insurance premium . . . yet.

Certification can also have a positive effect on staff morale. Notwithstanding the fact that staff may well be eager to make a contribution to bettering the environment through their work, installation of an EMS may not appear to them as particularly relevant to these goals. A certificate and the focus on the process that leads to its achievement can give a tangible goal to something that would otherwise have to be expressed purely in terms of figures on paper. Certainly at the very beginning of the EMS cycle, achieving certification is a valuable boost that comes at a time when results may not yet have been realized.

However, even these benefits would have to be weighed against the organization's knowledge of its own market. Is anyone actually demanding certification to the standard or the scheme as part of their trading process? Is it likely that the national governments of your export market will ask for certification in the near future? Is a regulator asking you to become certificated as part of their regulatory regime or as part of a penalty for a pollution incident? Is a major customer keen to support a phased implementation scheme as part of their own supply chain risk assessment process? Under these circumstances, the decision is a straightforward one, but without such obvious landmarks, an organization must take the temperature of its own market and act accordingly, taking into account the extra costs of achieving and maintaining the certification.

It is important that a certified company does not overstate the case even when it has achieved external assessment against a standard or scheme. Not all certification bodies are accredited for work in the same areas of industrial activity, and despite the best efforts of the accrediting bodies, some certifications are regarded as more equal than others. Neither does certification to ISO 14001 mean that the company is in full compliance with all the legal requirements all the time. Certificates can be withdrawn if there are persistent breaches of the regulations, but that doesn't mean that breaches won't occur.

In the same vein, certification to ISO 14001 or successful inspection against the achievement criteria of an intermediate stage in many phased implementation schemes does not make your organization a 'green' one, with an implied better level of environmental performance. It is the EMS that is being certified, not the output of the system. Of course it is difficult to separate outputs entirely from the system, as a breakdown in an EMS will lead to some non-conforming outputs, but it is still the system and its relative health that is being measured and assessed.

EMASs, on the other hand, along with certain other phased implementation schemes are said by some observers to have greater credibility in certain markets precisely because they include an element of environmental performance in the verification process. In the case of EMAS in particular, although it does not penalize companies for not achieving stated targets, the very fact that a public statement of performance is required ensures that companies who elect to use this scheme are certain they are going to be able to maintain their systems and their continually improving performance over a period of time.

Again, organizations will need to assess their own markets and select the most pertinent scheme for themselves. With greater congruency between ISO 14001 and EMAS, moving from one to the other is greatly eased. Phased implementation schemes can also accommodate companies who wish to go for one or the other or even both. Companies will be able to do this at their own pace, which not only befits the whole ethos of self-regulation, but also allows companies to get used to a new managerial discipline and series of tools before considering going public with a report on their performance in this area.

What can I expect from my inspector/certifier/registrar/verifier?
It is worth starting with the definition of the different roles of those involved in the certification or assessment process:

- Accreditation body. An organization that extends accreditation to certification bodies/registrars and/or verifiers, in order that they can ensure all assessment and certification is offered in a consistent manner across national and regional boundaries. Such a body is usually (though not always) a national one, but is always appointed by the national government for the purpose.
- Certification body/Registrar. This is an organization that is accredited to offer registration/certification against named standards and specifications, including phased implementation schemes.

- Verifier. An individual or organization that carries accreditation for the validation of an EMAS public statement and supporting management system.
- Inspector/Certifier/Assessor. An individual employed by a certification body/registrar in order to carry out individual assessments against the named standard or specification.

Unfortunately, many people who are not directly involved with the process of third party assessment use almost all the above terms as though they were interchangeable. Consequently, checking that everyone has the same understanding of your terminology is always a good idea.

At first sight it is hard to see why it is important to understand the way that external certification is organized, and how it applies to you and your organization's specific circumstances. The answer lies in the fact that accreditation bodies, certification bodies and assessors all have an influence on the way the requirements of the standards (or other specifications) are applied to your company. Although it is true that, say, the ISO 14001 Technical Committee 207 members are the only individuals who can 'interpret' the standard, all the above players will affect the final outcome of your assessment. As a consequence, where the standard or scheme you choose may be open to some broader readings, it is worth consulting the bodies and supporting documents that will be used as part of your own assessment.

Each of the certification bodies should be accredited to offer you their service, though it is worth checking to ensure that they are specifically accredited for phased implementation schemes, your own industrial activity, and for EMSs in general, as appropriate. If you are seeking to keep assessment time to a minimum, there is a strong argument to use the same certification body that you may currently use for your ISO 9000 registration, but don't make an assumption that your quality certifier has the accredited expertise in the environmental arena.

If you want to, you could also read the documentation that supports the accreditation of your chosen certification body, though remember that such documentation and guidance will necessarily be very broadly expressed, being, as it were, once removed from the specific assessment process involving your own organization. In particular you may wish to see guidance from the International Accreditation Forum (IAF), and/or the European Accreditation of Certification (EAC), both of which have documentation in this area. Your local standards supplier or your chosen certification body should be able to help you get hold of copies. An ISO committee on conformity assessment (ISO CASCO) have also revised ISO/IEC Guide 62, which originally only covered QMSs. Now published as ISO 17021, again, check with your local standards supplier for the latest document.

It is not only the accreditation bodies that influence the way that assessment is carried out. Certification bodies themselves may interpret different areas of the standard or scheme in different ways. It is worth checking with your chosen certification body at an early stage about certain areas that will have a bearing on your assessment; areas such as:

- Definition of 'site'. This could mean the physical boundaries; in certain schemes it may also include such satellite offices as distribution points, hub offices and the like.

- Policy towards multiple site assessment. Some certifiers are prepared to accept multiple sites on a sampling basis, where the certificate is issued prior to all the sites having been inspected. Even then, it is possible that all the sites will have to be inspected within a certain time limit, which may vary.

- Sites where there are shared facilities. Where services such as energy delivery and waste disposal routes are shared, some certifiers may have different policies concerning the amount of control they expect to be exerted. Check in advance.

- Scope of final certificate. What appears on the final certificate is extremely important, so it is worth checking the exact wording of the scope itself. Even within the bounds of their accreditation, certifiers have some flexibility here. The scope may apply to a specific address, or organization or even to certain parts of an organization. Check you are covered where you need to be, or you may need more than one certificate.

- Audit cycles. If you are choosing a different certifier from your ISO 9000 certifier, check that the audit cycles are the same, or consider moving your 9000 registration to another certifier in order to ensure that they are. (See below for a further discussion.)

- Incorporation of industry sector specific guidelines and initiatives. If you have used one of these guidelines, certifiers may choose to use such a document as a guideline but not formally recognize the contents. Check their attitude to whatever guidance you have used in advance.

Even within the requirements of their accreditation, certification bodies can have different policies in these areas, and some approaches may suit your particular circumstances better than others. If you have unusual circumstances that you think apply to your site or organization, or you are not sure of certain basic definitions, or if you need an audit team with particular expertise, remember to bring these factors up at an early stage of your selection process when comparing certification bodies. You may find that the choice is not as wide as you thought. Check too that they are EMAS verifiers if you wish to adopt that scheme at a later date or that they can offer a phased implementation scheme as an alternative for your organization or as an initiative for your suppliers.

Ask each certifier for a breakdown of how they run the assessment process, and ask particularly if they can carry out the work in multiple 'events'. This is not to be confused with a phased implementation scheme, but is a series of visits (normally two or three) that forms part of your overall EMS assessment. Even on the smallest site, it is worth having specific elements of your EMS checked first, to find out if there are fundamental flaws in the system before committing yourself to a full blown assessment. We recommend that you have the documentation, environmental aspects identification and evaluation procedure and the relevant regulatory identification procedures checked prior to the rest of the system. In our experience, a problem in one of these areas will simply be replicated throughout the rest of the system, limiting the effectiveness of the assessment feedback. If your certifier won't organize the assessment to your liking, see if you can find one that does, especially if it is your first experience of EMS assessment.

When you have had your assessment or inspection, our advice is also to read the assessor's findings very carefully. Certification bodies have to be very careful in the way they present these reports, lest the accreditation authorities judge that the reports veer too close to the field of consultancy. When an assessor discovers a non-conformity, they cannot advise you on the specific measures that you have to take in order to satisfy them, and hence the neutral wording of many of the assessor reports. However, most assessors are professional enough to word their reports in such a way that your corrective action should be in no doubt, so read it carefully when you are framing your response. When you have outlined how you intend to address the non-conformities noted during the assessment, the assessor will usually informally indicate whether your intended action is acceptable before you undertake any work.

For those undertaking the EMAS verification process, as it is known, it is fair to say that an EMAS verifier (either as an individual or as part of a team) has far more leeway in the way they express themselves. In particular, they are required to indicate to organizations what specific changes to the public statement need to be made that will make it acceptable to them. If you are having an ISO 14001 assessment and an EMAS verification at the same time, try not to be confused by the varying approaches of these different roles, even though they may be within the same team.

How is certification/registration/verification maintained?

Many certification bodies turn out the old cliché that 'Certification is a process not a target' and of course like most clichés, the truth that is at the heart of them can sometimes get obscured by their overuse as an expression. In the case of external certification to the EMS standards, this is particularly true, because losing an EMS certificate, or being de-registered for EMAS is perhaps potentially more damaging to the public reputation of a company than not achieving the standard in the first place.

One area to consider during the time immediately following the achievement of the certificate is the organization of your internal audit cycles, and how to get them to be in synchronization with the audit cycles of your certifier/verifier. By audit cycle, we mean here the time taken to run a complete audit of the entire EMS, including all the elements therein.

If you are maintaining a certified QMS, then you probably already know that many certifiers run a two year cycle, meaning that within two years, the certifier will have revisited the organization often enough to have considered the complete system. If there are four visits in that time, it means that the assessors must examine a different 25 per cent of the system each time they visit. If you are looking to make savings in terms of visiting assessors' fees, then it makes sense to ensure that your EMS is internally audited using the same cycle. There is no specification in ISO 14001 on this matter, but you may find that your national accreditation body has placed a requirement on all certification bodies to limit any extended cycles.

APPENDICES

In the case of EMAS, the matter is stated clearly in the current version of the regulation itself, where a maximum audit cycle of three years is allowed. It is also worth knowing that EMAS verifiers themselves have a high degree of flexibility in assigning the audit cycle for the scheme, though Annex II of the regulation and separately available verifier guidelines indicate that they must ensure that the chosen cycle is appropriate to the nature and scale of the organization and its environmental effects. Whatever the final audit cycle arrived at, one thing is clear: the strongest weapon that an environmental manager will have in maintaining an EMS is the organization's own audit programme and associated procedures. The danger period for most immature EMS is usually one year after the issuing of the certificate or the achievement of registration. The better an internal audit programme is during this period, and the more effective it is in seeking out non-conformities and improving the system, the better the chance that an external audit will find little of note when they revisit. Continual improvement sounds like a tall order under these circumstances, but it is possible, especially given the changing nature of any organization and the ability to pursue targets further down the supply chain as well as on individual sites.

Most certification bodies operate in a competitive marketplace themselves, and many offer some of the best customer service to be found anywhere in the world. As a result, it is possible to build up a good working relationship with such a body, but without endangering the independence that they will need to make the assessment services they offer credible. They should not be used as a substitute for employing an external consultant, however, for that very reason.

Finally, when it comes to seeking advice from external consultants on certification (or on any aspect of environmental work for that matter), be aware that every time an organization does so, it is passing up the chance to develop the expertise in-house and invest in the skills of its own workforce. There may well be other pressing reasons to employ consultants, such as restriction in terms of resources, time or perhaps even the complexity of the task and the quality of the finished product. Even so, it could be that such a short-term response creates a longer term problem. Many larger organizations have developed their own 'centres of excellence' in this area, allowing staff to be passed around within a corporate group or from site to site. For smaller companies, local or regional initiatives run by local authorities, or local offices of national government may have access to similar expertise.

In general it is worth ensuring that you have defined the brief for a consultant very tightly from the very beginning. The looser the brief, the longer a consultant will have to take in defining it for you, all the while charging you for the process. Ensure too that the consultant you use has sufficient professional qualifications in the area that you require. Professional bodies run registers of members and some even run referral services to help put you in touch with the specific level of expertise required. One thing is clear: don't assume that the skills required for installing ISO 9000 are the same for installing ISO 14001 or EMAS.

Appendix III
Environmental Management System Project Management

In its early stages the EMS is a project to initiate and manage. Later, it evolves and becomes part of the routine management in the organization. Although the EMS shares some common ground which we discuss in Appendix V, it is different from other initiatives such as quality and health and safety. Before examining the specifics of EMS implementation project management it is worth spending a little time to consider what we mean by a project.

Projects have some key characteristics which the environmental manager will need to know and understand if their own project, the implementation of an EMS, is to take root and grow to fruition in the organization. Perhaps the most obvious characteristic is that the EMS project has to achieve a specific purpose; that is to establish an EMS for the organization. Such an undertaking has eight key characteristics which signal that it is a project and not a routine activity that is part of the organization's normal business. As the EMS becomes embedded in the management structure of the organization it will become a routine operation and part of day-to-day management activity.

Table AIII.1 summarizes the key characteristics which distinguish a project from a routine operation. Although not all of these eight characteristics will reveal themselves at the initiation of the EMS project, they will during its implementation. Perhaps the primary characteristic of a project which the project manager needs to be aware of is that, above all else, they are the ones who will be judged on whether the project succeeds or fails. This book is designed to help the EMS project manager to succeed. The British Standard, *BS 6079-1:2002 – Guide to Project Management*, defines both a project and project management. The definitions are a useful summary of the terms:

> **A project is:** 'A unique set of coordinated activities, with definite starting and finishing points, undertaken by an individual or organization to meet specific objectives within defined schedule, costs and performance parameters'

> **Project management is:** 'The planning, monitoring and control of all aspects of a project and the motivation of all those involved in it to achieve the project objectives on time and to the specified cost, quality and performance'

APPENDICES

Table AIII.1 Key characteristics of a project

Project characteristic	EMS notes
It implies change	EMS undoubtedly represents 'business as unusual' with significant and positive change in the way the organization manages itself
It has a specific goal	In this case it is to actually install a business-focused EMS
It has a start and finish	Installing the EMS must have clear milestones in its planning. The most obvious are those that signify start and finish of the project
It has a specific practical outcome	The EMS is a living, dynamic creation with practical and tangible benefits for the business
It is original and unique	If you are installing an EMS it will be original and unique for your organization
Someone has responsibility for it	In the case of the EMS there will be an environmental management representative supported by a team of colleagues
It needs resources	As we discussed in Chapter 8, an EMS, like all other projects, requires human, physical and financial resources to make it happen
It needs diverse skills	It is unlikely that one person will have all the skills required for the EMS project, but that they do have the skill to recognize this and bridge skills gaps

These definitions raise three important areas – time, cost and quality – which need to be considered in the establishment of any project, and are particularly pertinent in terms of the developing EMS. These three aspects are often called the 'eternal triangle' of project management. Throughout the management of the EMS installation, the relationships between time, cost and quality must be regularly reviewed. There are often trade-offs to be resolved by the project manager who must be able to justify these to colleagues, the EMS implementation plan and management who 'own' the project. One example, from our Smallco case study, illustrates this point (see Figure AIII.1).

As the example illustrates, changes to one of the variables, for example halving the time to produce a register of environmental legislation may well save on costs by reducing staff time, but may result in an incomplete or interim register. Often having to repeat a task – in this case to produce a 'final' version of the register – may well more than double the cost as expressed in staff time. Beware cutting corners. If you do, then recognize the medium- or long-term consequences of such action.

The EMS project life cycle
It helps if the project has structure. Structure allows the project manager to design an effective implementation system and provides some clear milestones to measure and

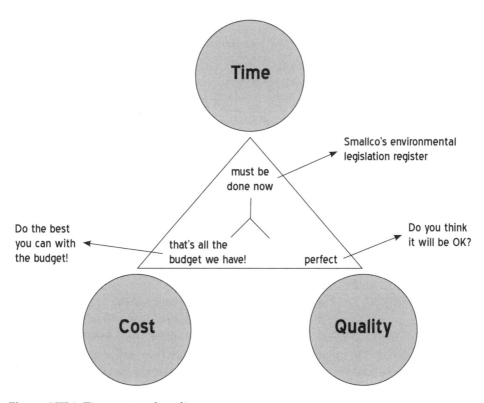

Figure AIII.1 Time, cost and quality

monitor progress. These are vital for the project team itself. If they aim at nothing then how will they know they have got there? The clarity provided by a project management structure will also enable considered reporting to take place at appropriate points in the process and keep senior management, who are likely to be resourcing the work, appraised of progress. A typical project will have seven distinct phases to it. Collectively these are often referred to as the 'project life cycle'. The list below summarizes them in the context of the EMS:

1   **Feasibility.** The pre-initiation phase typified by a report or study to decide whether EMS is relevant or not. See Chapters 1 and 2.

2   **Initiation.** The scoping of the project based on a thorough understanding of the issues and their implications. Initiation determines the terms of reference of the EMS: the foundations on which it will be built. See Chapters 2 and 3.

3   **Specification.** This is the determination of the 'whats' of the EMS. Here the detailed requirements of the EMS are agreed upon. This phase will require close collaboration with many internal and external stakeholders to ensure that the EMS meets the business needs. The understanding developed through detailed specification will enable professional judgements to be made about the likely resource requirements. See Chapters 3–6.

APPENDICES

4   **Design.** Having decided what, if anything, needs to be done, this phase determines how it will be done and engages a range of internal and external expertise. See Chapter 7.

5   **Build.** Having decided what will be done and how it will be done, this phase creates the systems to bring the EMS to life. See Chapters 4–6.

6   **Installation/implementation.** The EMS project is designed and built. This phase actually switches it on and it becomes live. Chapters 8–10.

7   **Operation.** Perhaps this is where, as all the previous phases are delivered, the project becomes integrated into the day-to-day operations of the organization and ceases to be a project at all. It is now an operational programme. As the EMS evolves, many new 'projects' will emerge and the life cycle starts up for each of them.

In reality, most of these phases are not really distinct at all but will blend into each other and overlap. Despite this blurring of the edges, the interfaces of each of these phases do present excellent opportunities for reporting on progress. They might be thought of as milestones in the EMS project process. As EMS project manager, helping colleagues to read the small print on each milestone is a key art of your job. Installing an EMS is a lengthy process. Milestones provide a measure of progress and focus which is accessible to your project team and to senior management. Clear communication is of paramount importance. It is unlikely that you can or want to do all the work required to install an EMS and so the actions of your team and colleagues will really make the difference between failure and success. They must be clear about what you expect of them and how their efforts will make the difference required to install an effective EMS.

In reality, implementing an EMS will involve coordinating numerous small projects, each of which is a component of the larger whole. This reality can be regarded as something of a relief to the hard-pressed environmental manager. By breaking down the whole into relevant component parts then the project to installing an EMS becomes more manageable.

For smaller organizations, perhaps with limited resources, a phased implementation scheme will contain a different model, such as the one in BS 8555. From a project management perspective the model represents an alternative user-friendly mechanism to develop an EMS. The individual phases usually culminate in 'achievement criteria' at the end of each task. When each criteria is met, the task is completed and the project moves on. Taken all together, the tasks build into a completed stage and the stages build into completed phases. Gathering up all the achievement criteria for a specific phase then forms the milestones for the project and the basis of an external inspection. See Appendix I for more details of how phased implementation schemes are structured.

Whichever route you organization chooses in the end, a useful way of working is to devise a project 'map' detailing all the component parts, their relationships to each other and to the EMS itself. One way of doing this is to work through the first three phases of project management in the list above and transfer all the mini projects and things to be done on to Post-it™ notes. An installation team session, armed with a collection of Post-it™ notes and with access to a clear wall space or set of flip-chart sheets is an excellent mechanism to build the project map. During the group discourse, Post-it™ notes can be moved around, interrelationships actually drawn on and highlighted and a real map created for the EMS installation. Give this process plenty of time. It is a real opportunity to explore and deal with doubt from fellow managers. A doubter won over is worth ten 'sheep' when it comes to actioning the EMS. Implementation scenarios can be followed through in the group by playing out the drama of 'what if . . .' and 'yes, I can see how that might work, but . . .'. The investment made here will pay back by not having to undo or adjust very much later on. This has measurable payback in terms of resource utilization.

Running the EMS – moving from project to programme

The list above sets out the seven phases of the project life cycle. Most of the acumen needed to address these phases is dealt with in other chapters of this book. Phases 6: Installation/ implementation and 7: Operation are the phases where the EMS is switched on, comes to life and evolves to become part of the management programmes within the organization. These phases mean that the EMS moves from planning to being proactive and it is worth expanding here in a little more detail. It is always good to have some kind of 'handover' ceremony, particularly if the installation team is to be disbanded and operational responsibilities have to be owned by mainstream management. Without a handover, many staff (including senior management) will think the implementation team are still the ones who will solve all environmental problems. Even if you decide to risk not marking the handover process in some way, the move from planning to practice involves attention to two key areas: EMS alignment and achieving objectives and targets.

EMS alignment and integration

To ensure compatibility, each of the elements of the EMS needs to be aligned and integrated with existing management systems. With some aspects of the EMS this is fairly clear. The environmental policy, EMPs, communication, training, accountability and responsibility, measuring and monitoring all need to complement existing management systems and not try and create new conflicting systems and a new bureaucracy with all its associated costs and negative connotations. Often, there are clear and well-established procedures in place which can be adopted and mirrored by the EMS. The models used in communicating and reporting financial results are well established and overlap into environmental management. Environmental auditing is one obvious example of borrowing from financial management.

Achieving objectives and targets

In Chapter 7 we considered the setting of business and environment-focused objectives and targets. Translating these into an EMP in operational terms is a major undertaking which will affect the entire organization. It is now time to integrate these into the business so that good intentions become good practice. This transformation involves eight work areas, summarized in Box AIII.1.

> ### Box AIII.1 Project management: Eight areas of work
>
> | | |
> |---|---|
> | 1 | Changing the culture |
> | 2 | Changing, adjusting or creating the management systems |
> | 3 | Changing the employee attitudes |
> | 4 | Defining new and extended responsibilities |
> | 5 | Assigning accountability |
> | 6 | Establishing organizational structures |
> | 7 | Establishing new information systems |
> | 8 | Defining new operational practices |

1 Changing the culture

An effective EMS implies significant culture change. It is very much a migration from some of the accepted and normal ways of doing things. However, as an opener to the management process it is unlikely to inspire or motivate many of the employees. The EMS manager needs to ensure that the EMS is implemented in ways which create a culture and climate to ensure the carefully crafted objectives are achieved. SMART implementation will ensure an alignment between people's personal objectives and those of the EMS. Positive culture change can happen gradually because of the installation of an EMS. This gradual evolution is much more conducive to a healthy EMS than trying to impose a new EMS culture on an unwilling or apprehensive workforce. Chapters 9 and 10 deal with the associated communication and training issues in more detail.

2 Changing, adjusting or creating the management systems

Part of your work to specify and design the EMS will indicate that existing management systems have to be adjusted or in the absence of any appropriate systems to adjust, that new systems need to be created. Phase 5 of the EMS life cycle above is very much about creating the right systems to bring the EMS to life. Whatever the reality is in your organization, new systems will take time and energy to become embedded in the organizational culture.

3 Changing the employee attitudes

It is a truism that 'people who don't know, vote no'. Part of the task of the initial environmental management project development is to change employee attitudes so that they are on board for the environmental management journey. Chapters 9 Communication and 10 Competence, training and awareness deal with these aspects in more detail.

4 Defining new and extended responsibilities

The reality of installing an EMS will be a world of limited resources, fixed deadlines, multiple roles and the intrusion of other work responsibilities into the well-laid plans to develop an EMS. Your work to set objectives and targets and some sort of mapping, such as that described in the section above on project life cycle will have revealed numerous inter-related key tasks which need to be actioned to bring the EMS to life in a systematic way.

Each task should be made the responsibility of one individual. Clearly, one individual may have many tasks to follow through from initiation to operation. The trap to avoid at all costs – remember time, cost and quality here – is having one task assigned to multiple individuals. Responsibility shared can often equate to a task not completed. Obviously one task may well require input from many individuals, but make sure that the responsibility for coordinating this lies with one person.

In defining and assigning new and extended responsibilities it is important to consider a number of factors:

1   Can the person actually do it? If not and there is no other reasonable choice then skills training may be required.
2   Is the person free to do it? Ignoring the question about why they were selected in the first place, what alternatives are there? A golden rule in terms of estimating the time required is to assume people are only productive for four out of five days.
3   What would happen if the assigned person left the organization or moved departments? How would this impact on the implementation of the EMS and how could you bridge any gaps that might appear?
4   More resource does not necessarily mean a quicker task completion.
5   External forces might force change not factored into the planning of the EMS. For example, in the UK, the Producer Responsibility Obligation (Packaging Waste) Regulations, which imposed procedural changes to organizations with an EMS and for those without one, forced new 'environmental' procedures to be created from scratch.

5 Assigning accountability

In terms of the EMS itself, the EMR is where the buck stops. They are the accountable person. Well perhaps not quite so. In the eyes of the law a board level director would be held to be accountable for a major and repeated breach of legislation. In a worst case scenario they would receive the custodial sentence. The EMR would probably simply lose their job! It is important that these boundaries are understood. If nothing else, a clear communication and understanding of them might be one way to ensure top management commitment for an effective EMS.

Internally, clear boundaries of responsibility and accountability need to be established. For example, an operative on the shop floor might be responsible for dealing with spent volatile organic solvent from a degreasing process, making sure it is disposed of in an agreed way and following approved procedures for the handling and storage of the solvent. That individual is responsible and accountable within the boundaries defined for their particular role. The next and important layer of accountability within the EMS is how these specific boundaries and procedures which define them have been derived. It is here that the activities on the shop floor actually interface with the EMS. Assuming that the solvent is deemed a significant environmental aspect of the business, then clear objectives and targets will need to have been set in the EMS to manage these aspects. This will be reflected in what actually happens in terms of observable practice in the workplace. The responsibility to deal with this, and a myriad of other issues, will have been delegated by senior management to the EMR, who has the authority to implement appropriate procedures. The final internal layer of accountability within the organization is senior management, for it is they who determine what the business actually does and how this is done.

6 Establishing organizational structures

While stating earlier in this appendix that it is not the intention of the EMS to create new and potentially burdensome bureaucracies, a recognizable organizational structure is needed to implement an EMS. An effective EMS does not operate in a vacuum. The EMS team itself might be, in part, a virtual organization with a few core players extended and complemented by other expertise as the need arises. However the team is constituted, formal reporting lines both upstream to senior management and downstream to other managers and the workforce are needed.

7 Establishing new information systems

Earlier chapters of this book have dealt with assessing the organization's environmental position and making business sense of it. This knowledge and understanding needs to be effectively integrated with how the business is managed. As we have discussed, installing and running an EMS will require many changes. As the EMS manager you will need to demonstrate that changes will make sense by quantifying the benefits. Take just one part of your work, perhaps on waste management. Table AIII.2 shows who might need to know about your work and what they might need to know. This is just one example of information needs that your EMS management will have to deal with. To satisfy such a diverse range of information requirements you will need to establish a rigorous, dynamic and ongoing information system.

8 Defining new operational practices

We have said that EMS is business as unusual and may well represent a departure from old ways of doing things. Analysis of aspects and impacts – of anything from products, raw materials, transport, energy, etc – and following this through to setting realistic objectives and targets means a new strand to the business planning process. It is not possible to plan and establish an EMS and miss this out. After all, the whole point is to ameliorate the significant environmental impacts of the business in ways which make business sense. These considerations will mean that new operational practices are needed. How these are actually defined is perhaps less important than who defines them. Once again, a multidisciplinary team approach will ensure a business-focused outcome in terms of new operational practices and, most importantly, an ownership by those staff who actually have to change the way they work. The role of the environmental manager is to create the climate for change and not attempt to impose their own – perhaps one-dimensional – solution on an unsuspecting and unwilling workforce. Chapter 10 suggests several strategies to facilitate such a climate for change.

Table AIII.2 Waste management information

Who needs to know?	What might they need to know about?
Production manager	Your analysis of volumes of waste being produced and the estimated cost. Recommendations for modifying aspects of the production process to reduce waste The benefits of changing practices, estimated and actual
Site manager	Impact on the site management of introducing skips for waste segregation. Responsibilities for managing any new arrangements including relationships with waste contractors.
Purchasing manager	Changes in purchasing patterns and sourcing of materials Suppliers' attitudes and involvement
Your regulators, for example, local environmental health officer or waste regulation officer	How you are recording waste management data under your duty of care regulations The monitoring data itself
External stakeholders such as insurers, financiers and the local community	That your are reducing the production of any hazardous wastes That you are sourcing new process materials, for example, replacing organic solvents with a detergent-based process

It is unreasonable to hope that the organization will change overnight or even over the weekend. Experience suggests that the changes indicated, many of which are decisive if the EMS is to take root, will need a long-term implementation plan with a life measured in years rather than months. Each of your environmental objectives and targets will have its own distinct requirements which will determine the shape and scope of implementation. As the environmental manager you need to have an excellent understanding of these aspects of implementation as they will signal the most effective means to integrate the individual components of the EMS and bring it to life.

What kind of information system to set up?

Whatever information system you set up it needs to cover the three 'A's: Accessible, Accurate and Adjustable.

Accessibility

It is worth spending a little time to jot down who you think will require access to the information and why they might need that access. Begin by reflecting on the IER in Chapter

3 and those colleagues both internally and externally with whom you talked. Ask yourself five questions and note your responses:

Q   Who have you already talked to?
Q   Who will you need to talk to?
Q   Who will want to talk to you?
Q   What might they need to know?
Q   What do I want them to know?

The short example on waste management in Table AIII.2 should give you some ideas here. Whatever you come up with from the work above, your information system needs to include:

✓   legislative requirements;
✓   health and safety requirements;
✓   records of existing and past consents and licences relating to emissions and discharges;
✓   a summary of your IER from Chapter 3.

Accuracy

Work on the IER in Chapter 3 – to collect the environmental data, extending this through aspects and impacts, a meaningful policy and finally setting realistic objectives and targets – now needs to be accurately integrated into your new environmental information system. Your measurements will have begun to quantify your environmental performance and, at the very least, provide a baseline to measure progress against. Accuracy is a vital component in substantiating the validity of your work. Consider the impact of these two statements if you were addressing a transport department:

- I think we should be reducing our transport costs because we'll save money and help the environment.
- Last year our energy costs were £x on diesel and £y on petrol. I have calculated that by devising a new delivery routing, training drivers in fuel efficient driving at a cost of £z we would save 8 per cent on our total fuel bills and pay back the investment in under six months.

Accuracy is the key to persuasive argument and action. Your information system needs to be accurate, objective and specific.

Adjustment

Environmental management is a dynamic process. Legislation regularly changes, new, low impact technologies and techniques emerge, staff are recruited or move on, product and process change over time and who can say whether you will still be in the same business five years down the road? All these changes, many of which are outside your control, will happen to some degree or other. Your information system needs to be robust enough to deal with these changes. The third information design requirement is to make sure that your system can be adjusted to cope with change.

Ways to set up your system

Your information system needs to acknowledge the ways the information will be used and the ways you would like it to be used. One way to design the system is in two related parts:

1   The hard statistical data; perhaps measurements of energy, water, waste and emissions and discharges. Although the actual measurement and monitoring might involve paper recording, it makes sense to go electronic as a way to ensure the data are quickly on hand in a form or forms that they are needed. Most computer databases can be modified to suit your needs and there are a growing number of dedicated software packages appearing on the market. Whatever route you choose you need to be in control. Make sure the system meets your information storage needs and can handle your raw data.

2   The paper; some data, from your stock take and process measurements, you will want to have at hand. Some of your information will be needed in an easily retrieved paper form and remember that some data, relating to regulated processes, need long-term storage and retrieval systems. Set up an indexed, paper-based system to handle these data and lodge copies with relevant staff who need regular access to it.

For both your computer and paper-based information you should make sure that:

1   You set up maintenance systems to keep the information up to date. It may make sense to delegate some of the responsibilities to staff concerned more directly with the issue.

2   You indicate who collected the data, when they were collected and how they were collected.

3   You set out how the data were analysed and any actions arising from this analysis.

4   You note cross links with, for example, your health and safety policy and practices.

5   You note how long the data must be kept.

6   You note the level of confidentiality of the data and who has access to them and for what purposes.

Chapter 11 Documents and their control, deals in more detail with these issues.

Appendix IV
Planning for Continual Improvement

If you aim at nothing, you'll hit it every time (Steve Van Matre, 1990).

So you have an EMS in place. What happens next? This short appendix suggests a rationale and some ideas for keeping things moving.

If your system stagnates, then the people involved in its day-to-day running will grow stale too. On its own, an EMS which might be registered to ISO 14001 or EMAS is simply the minimum an organization can do. The spirit of both standards, and in particular ISO 14001, is to do more than simply comply with the clauses and be able to tick them off one by one. ISO 14001 calls this ongoing evolutionary process 'continual improvement' and signals it in the form of an upward spiral in the systems diagram of the standard (see Figure AIV.1). In ISO 14001 continual improvement is explained as a recurring process designed to achieve continuing improvements in environmental performance which are consistent with the organization's policy. It is therefore a key attribute of an effective EMS. We believe that if continual improvement is to be sustainable then it must be integrated with commercial improvement and continued engagement of top management. Your management review (see Chapter 14) is the perfect place to formalize your commitment to continual improvement by taking time to consider how you can push the boundaries of your system in ways which make sound business sense. The wise environmental manager will be well prepared to use the management review as a launch pad for such improvements.

Going beyond the minimum requirements is really what continual improvement is all about. It is a dynamic process which constantly seeks to identify ways to improve the EMS to further reduce pollution and in ways which make sense for the business. The management system standards provide this scope to allow the EMS to evolve so that it continues to deliver these twin goals for the environment and the business. Perhaps at first, the installation of an EMS may represent a challenging responsibility for the environmental manager and their team. Over time, the EMS will become a more mainstream part of the organizational culture and a regular programme of review will provide acceptable recommendations for changes to the system. It is likely as the EMS matures, that minimum adherence to one of the standards may not provide adequately for the needs of the organization.

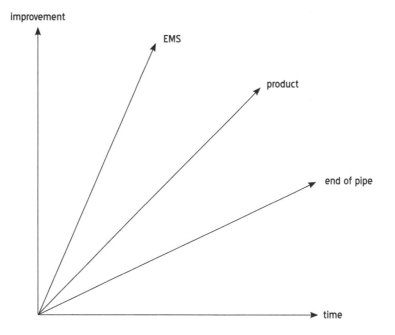

Figure AIV.1 Delivering improvement

Remember, we have repeatedly asserted that an effective EMS is a live organism. Survival of any organism depends on its ability to maintain its position in the scheme of things. If the EMS is something the business wants to conserve rather than preserve then work will need to be done to look after it. Continual improvement is a really a twin strategy to polish and tune the EMS to ensure that it continues to work both for the business and the environment.

Strand 1 Improving environmental performance

In Chapter 3 we suggested that an IER would help provide the baseline data for the developing EMS. Linked to this was the analysis of environmental aspects and impacts: the cause and effect of your businesses environmental issues. The final linkage, made in Chapter 5, was to identify and satisfy legal and other requirements. These three areas set the scene which allowed an environmental policy to be drafted and meaningful objectives and targets to be established. Your work to bring this intent to life through management programmes, and measuring and monitoring your progress towards these targets will have defined the unique nature of your EMS.

Improving environmental performance is an uncomplicated approach and builds from this implementation work through your analysis of how well you are doing; in this case the data for the analysis are gathered during environmental auditing. The ongoing review of your EMS provides guidance on how to improve environmental performance. There are several key questions you need to ask yourself about how well you are doing. This process of analysis can be summarized using Table AIV.1.

Table AIV.1 Analysing environmental performance

Objective: (name of objective here)				
Targets	Target reached?	Corrective action? or Preventive action?	Remedial action(s)	Opportunities for improvement
1				
2				
3				
4				
5				
etc				

For each action taken or planned there are three questions to ask yourself:

Q1   Is this something we must do? For example, to stay within the applicable legislation or to ensure established waste reduction targets are met.

Q2   Is this something we will do? For example, to change procedures to save raw materials.

Q3   Is this something we could do? For example, investment in new plant which would reduce pollution and lead to a business return on investment.

Strand 2 Improving business performance

If the concept of continual improvement is to take root in the organization it needs to be more than simply improving environmental performance. The second strand and – legal compliance aside – the one which will move good intentions into good practice is to ensure that whatever you do will result in bottom line benefits to the business. Most of the examples we have given in the book illustrate that the twin goals of improved environment and business performance will be achieved by an effective EMS. A vibrant EMS will ensure that these twin strands are delivered throughout. The second strand can be addressed using cost benefit analysis (CBA). There are many techniques to carry out CBA from simple accounting methods to detailed costings linked to an organizational model much like the life cycle analysis discussed in Chapter 4. Something in between is suggested here to help begin the process of identifying environmental management actions which will improve the business performance.

Cost benefit analysis: A ten step approach

1   Describe the issue.

2   Consider stakeholder links: the law; finance and insurance; customers and suppliers; staff; the community; good management.

3   SWOT the issue: analyse the Strengths, Weaknesses, Opportunities and Threats.

4   List potential disbenefits.

5   List potential benefits:

       (a)   what must you do?

       (b)   what will you do?

       (c)   what could you do?

6   Consider investments and returns.

	TIME		....	
INVESTMENT STAFF	HOW	....	RETURN
£/$	LONG?	....	

7   Estimate total return(s).

8   Consider the decision areas, for example, purchasing, procedural changes, early success.

9   Provide supporting information, for example, best practice; case histories and industry/trade association guidelines.

10   Communicate clearly and get a decision.

Remember, as we have emphasized throughout the book, don't go it alone with your CBA. The work is an excellent opportunity for involving colleagues to share in the analysis of success.

Putting it all together: Windows of opportunity

A consideration of improvements you could make to environmental and business performance will have revealed numerous routes towards continual improvement. Clearly you can't do it all at once. The standard ISO 14001 acknowledges this by indicating that continual improvement doesn't need to take place in all areas simultaneously. Business reality also advises against work on all fronts simultaneously. Not only might the environmental manager become exhausted, but the staff who are the organization will quickly switch off if environmental issues threaten to dominate their working lives. What is needed is a prioritization of continual improvement opportunities so that the focus of your efforts will be on those areas that will make a difference to environment and business performance. This can be represented by considering the windows of opportunity that exist in your organization and opening those windows which will illuminate and extend your EMS.

Table AIV.2 shows a simplified 'window map'. It suggests multiple individual 'windows' or areas for improvement. For your own organization, extend the map to generate ideas which meet your requirements. This illustration, for Smallco, identifies sample opportunities out of a possible 180 (ie 12 drivers × 15 issues). The final step is to actually 'open the window'. Remember the Rudyard Kipling guidance on how to do this in his poem to young journalists:

Table AIV.2 Window map

Environmental issue area objective/target (see Chapters 3 and 7)	Regulators	Employees	Business drivers (see Chapter 2)				
			Unions	Shareholders	Bank	etc	
Management issues	Ensure confidence among senior management	Greater involvement, motivation and productivity	Excellent 'common ground' for positive relations	Impress shareholders with proactive stance	Impress bank with proactive stance	
Legislation	Guidance on how to keep up to date	Confidence that employee cares	No negative agenda items re legal issues	Confidence	?	
Marketplace	Customers need to see compliance plus	Improved external relationships	?	Better balance sheet	Objective evidence for business plan re loan	
Distribution and transport	Effective policy to ensure compliance	?	Improved relationships and communication	?	Confidence in management	
and so on . . .						

> I keep six honest serving men
>> (They taught me all I knew)
> Their names are What and Why and When
>> And How and Where and Who
> *(I Keep Six Honest Serving Men*, Rudyard Kipling*)*

For each 'window' the Who, What, When, Where, How, Why questions provide the keys to open it and implement an opportunity for continual improvement which will bring environment and business benefits. Your responses to the six questions provide guidance on how to implement actions both to improve your EMS and to justify these actions to staff, from senior management to shop-floor operatives.

Q1 What will you actually do?

A1 Your analysis of environmental and business opportunities will signpost 'whats'.

Q2 Why will you do it?

A2 Your analysis of environmental and business benefits will tell you why.

Q3 When will you do it?

A3 Your integration of actions with both the EMS and the business will identify an implementation timetable.

Q4 How will you do it?

A4 Remember SMART from Chapter 7. Your action plan for continual improvement will need to follow these guidelines.

Q5 Where will the action(s) take place?

A5 This could range from shop-floor interventions involving procedural and production changes to training and communication interventions both internal and external.

Q6 Who will do it?

A6 Although the driver is likely to be yourself as the environmental manager, the actual work may well be delegated as appropriate.

Don't forget to keep your windows 'polished'. This will involve monitoring and measuring and the regular review process inherent in any good management system. Over time, as your EMS evolves and matures, not only might you be guided by best practice, you might even be establishing it too.

Continual improvement might be mistaken for being stuck on the up escalator with perhaps a worry that you will eventually hit the glass ceiling. Don't worry; there is always plenty of headroom. As your EMS matures and you have developed an effective management system to deal with internal issues then you can begin to direct your efforts outside the organizational boundaries. Remember the life cycle analysis in Chapter 4. This approach suggested the external links that exist for any organization. As you become more confident with your own EMS you can begin to exercise control and influence up and down your supply chains (see Figure AIV.2). This may even be where you started with your EMS: on the receiving end of a customer's questions about your own environmental practice. You might even be asked to join an industry working group tasked to devise the best environmental practice advice for your industry sector.

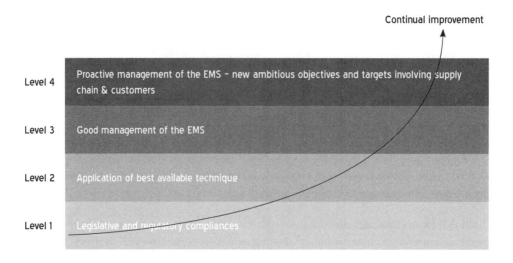

Figure AIV.2 Environmental management systems evolution towards continual improvement

Appendix V
Integrating Management Systems

EMS do not operate in a separate silo from other business systems. Many readers of this book will already have others in place as an accepted part of the business culture. These might already include quality, health and safety and perhaps some strands of management that apply to corporate responsibility or even sustainability.

The impact of QMS has been growing steadily since the publication of the British Standards 5750 – *Quality Management Systems*, in 1982; now superseded by the international standard series, ISO 9000, the vocabulary of quality has become an accepted part of day-to-day management. Quality management standards indicate best practice as defined by international experts in the field and can be used to provide assurance of an organization's ability to meet defined objectives. Closely linked to quality management and the standards that define it, is the concept of 'Total Quality Management' or TQM. TQM is a change mechanism which is targeted at developing a customer driven organization through integrative working practices. The underlying purpose of TQM has much common ground with an EMS though the modes of delivery can be quite different.

Health and safety management can be regarded as a set of external controls dictated to business and which are achieved by establishing and fulfilling specific targets. In some countries the law requires an organization employing more than five people to have a well defined health and safety policy, a trained and proficient member of staff to implement the policy and to carry out the necessary risk assessments of operations and to develop related control activities. Organizations which effectively manage health and safety exhibit some common characteristics. They have their health and safety risks under control and strive towards continual improvement in their health and safety record. Again, several echoes of environmental management, but more at an operational than strategic level.

These systems, quality, health and safety and now a growing requirement to address environmental aspects of the organization, coupled with social responsibility, governance and sustainability, all represent external pressures for change. For example, a key customer might require all suppliers to achieve ISO 14001 within a defined time period. The outcome of simply responding to these pressures as they arise, by reactive management, will result in an incomplete and inefficient outcome and miss the opportunities of what might be achieved through proactivity. Of course, as we have discussed in the opening chapters of this book, environmental management itself represents a similar change agent. Whether it results in actual business benefits rather than business burden will depend on the way implementation takes place.

In all management systems, no matter what the object of management, there are enough similar drivers, mechanisms and requirements to suggest that a common approach makes the best sense in terms of resource utilization. By integrating the management systems the business benefits can be multiplied and an organization can respond to change and anticipate or move beyond any targets, for example, changes in legal requirements, well before they are externally imposed.

> Effective health, safety (and environmental management) is not 'common sense' but is based on a common understanding of risks and how to control them brought about through good management (Successful H&S Management, *UK Health and Safety Executive, 1991*).

And that's the whole point – a risk is a risk is a risk, whether it's environmental, social or economic and everyone can understand managing risks. In this way, environmental management can provide the paradigm to turn strategy into action. A useful starting point in developing an integrated system is to investigate common ground shared between systems, as illustrated in Table AV.1. The ISO 9000 series gradually becomes increasingly customer oriented over time and now places greater emphasis on setting measurable objectives and targets and on measuring product and process performance. In short it follows a systemic approach to management characterised by ISO 14001. Approaches to developing a health and safety management system use similar models and even the development of sustainability management models bear evidence of the same PDCA lineage.

There are two key drivers which justify a move to a more integrated management approach:

1   Legislation. There is no choice here. Ignoring environmental legislation will eventually lead to prosecution, fines and ultimately could lead to imprisonment. This is equally true for health and safety management which must address both the 1972 Health and Safety at Work Act and the 1992 management of Health and Safety at Work Regulations. The legislative links with quality management are a little less concrete, but do apply to product fitness for purpose and health and safety links associated with product use. The bottom line is that employers have a responsibility to protect their staff from any risks associated with their operations. Quality is constrained (or encouraged) by product legislation imposing minimum expectations, while sustainability and CSR issues are figuring more and more in mainstream business legislation around the world.

2   Financial. Environmental, health and safety, or quality incidents will affect the bottom line. Fines for breaches of legislation are one obvious example of a fiscal concern, others are discussed in Chapter 5. The financial downside of not managing social responsibility issues effectively

Table AV.1 Environment, quality and health and safety

Key element	Environment	Quality (ISO 9001-2000)	Health and safety (BS 8800)
Top management commitment	Defined	Defined	Defined
Policy	Policy required	Policy required	Policy required
Planning	Aspects and impacts Legal and other requirements Objectives and targets EMPs	Quality planning Provision of adequate resources Objectives and targets Quality management programmes and admin.	Risk assessment Risk control Objectives Performance standards
Implementation	Structure and responsibility Training... Communication Document control Operational control Emergency preparedness and response	Management responsibility Product realization Training... Communication Design and development Operational control	Risk control Hazards documentation Document control Accident procedures Site rules Performance management techniques
Measuring performance	Monitoring and measurement Corrective and preventive actions Records	Measurement, analysis and improvement Planning Analysis of data Control of non-conformity Improvement – Corrective and preventive action	Performance standards Premises, plant and equipment People, procedures and systems Reactive monitoring (corrective actions) Active monitoring (preventive actions)
Auditing	EMS audit	Internal audit	Regular and systematic reviews and independent audits Commitment to continuous improvement
Reviewing and improving performance	Top management review of entire system	Management review	Management review to improve performance and further develop health and safety practice
Continual improvement	Commitment to continual improvement	Commitment to continual improvement	Commitment to continuous improvement

APPENDICES

is realized in an even more spectacular way through the loss of reputation. Meanwhile, although the financial penalties of being unsustainable might not appear very obvious at first, it only takes a second to realize that getting investment for unsustainable practice is pretty hard to do.

As we discussed in Chapter 2, environmental management has brought to the fore several new external drivers which reinforce the logic of moving to integration. These range from customers who have expectations and demands for improved environmental performance to society in the widest sense and increasing intolerance for poor environmental performance.

All management systems, whether externally defined or developed by the organization itself, require complementary and overlapping systems of operation. Action by many organizations has been to develop a '3 Ps' approach: Policy, Practice and Personnel. If taken in isolation from one another the resulting duplication of effort can be significant and will have cost implications for the organization, with each management system having its own, independent 3 Ps. The multiple systems have an added disadvantage of needing more complex communication processes to ensure that they complement rather than conflict with each other. Running separate systems can easily mean a hidden administrative burden on the business.

Integration at its best will reduce the burden of implementation and increase organizational benefits. Integration can magnify the strengths of each system while the negative aspects are minimized or even eliminated; for example, environmental management training can extend system boundaries and help the change process. The key positive and negative aspects of each system are summarized in Table AV.2.

Table AV.2 Positive and negative aspects of management systems

System	Positive	Negative
Environmental management	The dynamic and changing nature of environmental management means updating is an essential component of management Much more proactive than either of the other systems	The external drivers over and above legislation still developing, eg supply chain, new markets, customer demands Some still believe the issue will go away if they ignore it and behave accordingly
Quality management	Driven by the needs of the company and tends to be proactive Training is often founded on organization functions	Systems can be perceived as over complex and bureaucratic Compliance with external customer demands may simply result in a reductionist position
Health and safety	Legislation driven Mature policy and training systems often in place	Health and safety tends to stop at the factory gate

Table AV.3 Key issues

	Question	Environmental	Quality	Health and safety
1	Does your organization meet applicable environmental regulations and other requirements			
2	Does your organization comply with health and safety regulations?			
3	Does your business have to meet customer quality requirements?			
4	Do you have a written policy in any of the three areas and is it available?			
5	Does the policy:			
	• acknowledge internal and external drivers?			
	• identify hazards?			
	• assess risks?			
	• define control measures?			
	• set out responsibilities?			
	• require a training plan?			
6	Do you have procedures to monitor policy in action?			
7	Do you keep adequate records?			
8	Do you measure and monitor?			
9	Do you take action to correct any non-conformity?			
10	Do you have good relationships with your regulators?			
11	How much duplication of effort is there?			
12	When it comes to management policies, procedures and personnel (the 3 Ps), have you integrated (3 Ps) or kept them separate (9 Ps)?			

It is perhaps worth reminding yourself about some of the key issues implicit in the management systems using Table AV.3. Answer 'Yes' or 'No/Not sure' for each question. If you have answered 'No/Not sure' then look again at the sources of help and advice you have, both internal and external. For 'Yes' answers, review the information you have about each issue, and consider how an integrated approach may best be developed.

Implementing a joint management 'system' requires that the strategies which have been discussed throughout this book are brought together into an integrated system. Four areas need careful consideration:

1   required action;
2   combined monitoring progress;
3   a combined management manual;
4   continual improvement.

Required action

Many of the strategies described in the book are characterized by action: collecting information, making the relationships internally and externally, preliminary communications and more detailed considerations of a communication and competence and training strategy. Alone, this wealth of data and networking is worth little to the organization. An action plan for implementing a combined management system will draw together key findings and conclusions to tease out exactly what action steps will be taken to realize the aggregated benefits. There are three areas to think about to develop an action plan:

1   What needs to be done?
2   Objectives and targets.
3   Implementation.

What needs to be done?

The IER will have identified the key areas needing management action for the environment. This basic review system can be used to gather information about both health and safety, quality, social responsibility and sustainability management issues for the organization.

Objectives and targets

Set objectives and targets based on what makes sense and this will advise actions. Consider the priorities for combined systems in three action or priority areas:

Priority 1   Must do activities. For example to comply with existing legislation or customer requirements or those which can bring immediate wins with cost savings.

Priority 2   Actions which do make organizational sense but are not so pressing, for example, those bringing medium-term cost savings or those which satisfy anticipated customer needs.

Priority 3   Actions which will require longer term planning and investment.

A realistic timetable to start, carry out and finish each action is needed alongside these priority areas. Trying to implement any management system 'all at once' is a recipe for frustration and disintegration. Although equipment, process or systems changes might be necessary, as we discussed earlier, positive actions of staff are the foundation of sustained

Table AV.4 Outline action plan

Area	Target	Timetable	Key players
Energy	Reduce consumption • heating by 7% • machines by 5% • electricity by 10%	 • May 20?? • winter 20?? • March 20??	Staff/site manager Production manager Operations staff
Health and safety	Satisfy any outstanding legal regulations Maintain legal compliance	immediately end 20??	Health and safety manager Senior management
Waste	Reduce waste costs/ increase income • segregate scrap metals • improve purchasing specifications • reduce office wastes	 • from new skip lease period • at supply contract negotiations • November 20??	Site manager/contractors/ waste regulators Buyers/suppliers Office personnel
Product and process	Investigate sources of raw materials Cost new machines and investigate environmental issues	Spring 20?? End 20??	Purchasing manager Plant manager Finance manager

change. It is much more realistic to break down your implementation strategy into a series of smaller steps which can be achieved. An outline action plan might look like the example in Table AV.4.

This process of considering 'area – target – timetable – key players' should be considered for each issue for the organization. The issues that arise can be ranked as they are considered. Now consider the issues of your organization and construct a draft action plan using the four headings: 'area – target – timetable – key players'.

Implementation

Each particular ranked goal will require its own set of actions. This can be considered the implementation stage of an action plan. It has seven phases, the detail of which will vary from issue to issue. An example of an implementation plan is given below.

Phase 1: The data

Look at the total information for the organization:
• Where is the greatest impact: environment, CSR, sustainability, quality and health and safety?

- what are the causes?
- what could be done to reduce impact/costs?
- what could be done to improve the situation by being proactive?

Phase 2: The players

Who are the key staff involved: internal, contract and external?

- what are their feelings and ideas?
- Can they actually and positively contribute and if so how best?

Phase 3: The targets

Reach consensus on what the targets are to satisfy legislation and reduce impacts and costs.

Phase 4: The plan

Agree a plan of campaign with the key staff involved. Include:

- the proposed actions and changes;
- the responsibilities and the timeframe;
- resourcing required.

Phase 5: Ownership

Everyone involved – either directly or indirectly – must be fully committed and back the plan. This must include senior management throughout the process. Training and communication work will reveal complementary activities here.

Phase 6: Do it

Having reached consensus make sure the plan is carried out.

Phase 7: Monitor progress

The process will need to be managed and monitored, fine tuning each of the phases as necessary.

Monitoring progress

The requirements for a combined management system should now be clear and it is important to verify that what should be happening is actually happening. Let's take an environmental example that illustrates the same basic approach for all the different types of issues. In collaboration with one of the regulators your work might have established a new action on a legal requirement pertinent to your operations. If pollutants are emitted which exceed the agreed levels then not only could human health be jeopardized, but penalties could be placed on the organization. Effective monitoring and, if required, corrective action will prevent such damaging occurrences. Your progress needs to be monitored on two levels:

Level 1: The micro level

This is the monitoring of progress towards each goal in an action plan and the fine tuning of the process as knowledge increases over time. It is vital to establish a regular monitoring programme as a cornerstone of an integrated management action plan. There are a number of computer-based information management systems which can help deal with the growth in data and help to make sense of it. These can be set up to provide prompts to regularly monitor progress.

Level 2: The macro level

Your extension of the baseline review to encompass wider issues posed by CSR, sustainability, health and safety and quality will provide data for an integrated management system. Regular auditing to find out how well you are performing is vital.

On both levels, even if an organization is performing well from a financial perspective, managers still carry out financial audits to verify the results and to prevent any shocks from occurring. Exactly the same reasoning needs to be applied to the verification of your performance in terms of the other issues. It may be even more important, for example, a breach of health, safety and environment may have a devastating effect on the organization either directly through fines and legal action or indirectly through loss of well-established reputation.

A combined management manual

A manual can set out established procedures for all types of separate issues, not just environment, quality and health and safety management. It can be developed and broadened gradually to provide the key point of reference for effective management. Even if an organization is relatively simple in structure and undertakes few complex operations, the management issues faced are still multitudinous and cross-referential. A manual should become the point of reference for the organization and most importantly, for those staff with day-to-day responsibility for implementation of however many management systems you choose to develop and integrate. To be effective it needs to gather users not dust and should be designed to be accessible, not just to take up shelf space! It does not necessarily have to be paper based. It could be an electronic on-line system with controlled access so that those that need to use it can extract what they require and add comments and suggestions for the management representative(s) who looks after it. To be effective any management manual needs to:

- be clear and simple;
- be written in straightforward not technical language;
- be as short and accessible as possible;
- use charts and diagrams to bring the information to life;
- be dynamic and easily updated;
- be available to all who need access to it.

As with any new initiative, it makes good sense to produce a draft first and present this to colleagues. Some sections of the manual will relate to relationships with regulators. Assuming there are no issues of commercial confidentiality, then a dialogue with regulators at this drafting stage will improve the final product. A checklist for an integrated management manual is provided in Box AV.1.

Once an executive summary has been added, the draft manual can be presented and circulated for comment. Feedback from colleagues will allow the final version to be produced and the manual formally released within the organization as part of the communication and training processes developed earlier.

Box AV.1 Integrated management manual checklist

Purpose of the manual and a description of your management systems

Your policies for environment, quality and health and safety

Issue specific actions for:

- Energy
- Water
- Raw materials
- Process and product
- Site
- Paper and packaging
- Discharges
- Transport
- Contractors
- Community

Your implementation plans for each issue including targets and time-scales
Roles and responsibilities: the key players
Legislative requirements
Communication process and programme
Training process and programme
Monitoring/auditing process and programme

Appendices: Would also include related procedures and work instructions
Data for each issue
Related policy documents and plans
Computer record systems
Regulations and permits
Notes on management committee meetings
Incident reports and emergency plans

Continual improvement

Effective management is a continuous process, not a finite programme. This appendix will have helped establish a map of the territory and gathered the tools needed for the journey. The process should be well in place and become simply another component of good management. Effective monitoring will begin to realize tangible benefits both internally, through clearer business strategies, cost savings and staff morale, and externally, through improved relationships with customers, suppliers and the regulators. For any organization, this may be enough in itself. The implementation of effective management is more than just an obligation or expense. It will provide effective risk management, cost savings, increased profits and market share. Effective management always makes organizational sense. If it isn't effective, then management consideration is needed.

Perhaps the ultimate goal is to create a healthy, high quality organization with minimum negative environmental, social and economic impacts. The benefit of effective training will be an important contribution towards developing a healthy culture within the organization. Most of the benefits can be quantified and as we discussed in the Introduction, the best organizations are already turning benefit into business. Some measures of organizational health are listed below. How many would you be able to say yes to?

Less time lost through health-related absenteeism	Yes/No
Reduced occurrence of health and safety; quality and environmental incidents	Yes/No
Improved morale	Yes/No
A positive attitude to training and taking action	Yes/No
Fewer grievances	Yes/No
Greater loyalty	Yes/No
Improved timekeeping	Yes/No
Lower staff turnover	Yes/No
Recruitment of excellent staff	Yes/No
Less time lost due to workplace disputes	Yes/No
An adaptive culture and flexible workforce	Yes/No
Held in high regard by the local community	Yes/No
Fewer/no defective products delivered	Yes/No
Increased output and productivity	Yes/No
Lower level of machine downtime	Yes/No
More new business	Yes/No
Reduced customer complaints	Yes/No
Higher sales	Yes/No

All of these are outcomes of an effective integrated management system. All have a cost attached to them and all make the difference between success and failure for an organization.

Some final thoughts

- Most people will feel more comfortable integrating quality, health and safety and environment at first, mainly due to the tangible nature of the operational issues. Don't, however, be put off from looking at seemingly less straightforward issues. Corporate responsibility is merely what it says – the ability of an organization to respond to a problem. Systems help that response but they are not a replacement for strategic thinking.
- We are not suggesting that issues such as sustainability or social responsibility can be managed by a system alone. But a strategy for any of these issues is useless without a mechanism to deliver it. The development of a sustainability strategy is a completely different ball-game from designing systems that can turn the thoughts into action – in fact it's the

subject of another book and there are plenty of guidelines out there to help.

- What we are saying is that EMSs are used to coping with complex operational issues and they form a good model to help make things happen when the issues might seem difficult to grasp. They also help to set up an ongoing internal and external dialogue; a management system forms a bridge between strategic direction and operational feedback. Organizations need it now more than ever.

- Beware issues that might be 'lost' in any overlaps of responsibility. For example, the different areas will have different legal requirements. An integrated system must still designate responsibility and authority for management and not open up new cracks for things to fall into.

- Refer to customer and other stakeholder needs throughout the process. Keeping stakeholders satisfied is part of your licence to operate so an ongoing dialogue needs to be built into everything you contemplate undertaking.

- Manage audit trails carefully. These are likely to cross system boundaries. Careful audit planning is essential. Remember you are only auditing the system's ability to deliver – strategic direction is informed by the feedback but is not replaced by it.

- You are unlikely to be starting with a blank slate. It is vital to acknowledge and plan for this by carefully mapping together each area gradually over time: social responsibility, governance, sustainability, quality, health and safety and environment.

- Although integration will make more effective use of resources, beware the simple maths of dividing the implementation budget by three! The savings can be significant, but don't over-represent them to an eager finance colleague.

- If it doesn't make sense to integrate now then don't. However, do make sure you have an open and continuous dialogue with your colleagues to share ideas and understanding of what each of you are doing.

Glossary

Accreditation
Procedure by which an authoritative body formally recognizes that a body or person is competent to carry out specific tasks, such as offering certification or training.

Agenda 21
The global action plan developed at the Earth Summit in Rio de Janeiro in 1992. It attempts to balance future developments with environmental imperatives. The Rio Declaration identifies local action as the key to success for Agenda 21.

Applicant
Legal entity applying for an environmental label for a product or a range of products and that undertakes the compliance with ecological and product function criteria and the certification and costs involved in the application and awarding of the label.

Assessment
A judgement or determination of the significance, importance or value of something. In this case the EMS against the requirements of a specification or a standard.

Assessment body
Third party which assesses products, goods or services and may hold successful applicants on a register of certified suppliers.

Assessment system
Procedural and managerial protocol for carrying out an assessment which may lead to certification. This 'system' will also apply to ongoing maintenance.

Audit: general
A planned, independent and documented assessment to determine whether agreed upon requirements are being satisfied.

Audit: ISO 14001
A systematic and documented verification process of objectively obtaining and evaluating audit evidence to determine whether an organization's EMS conforms to EMS audit criteria, and communicating the results of this process to the auditee.

Audit criteria
Policies, practices, procedures or requirements against which the auditor compares collected audit evidence about the subject matter.

Audit findings
Result of the evaluation of the collected audit evidence compared against the agreed audit criteria.

Audit programme
The organizational structure, commitment and documented methods used to plan and perform the audit.

Audit team
Group of auditors or a single auditor, designated to perform a given audit; the audit team may also include technical and legal expertise.

Auditee
Organization to be audited.

Auditor
Person qualified to perform environmental audits.

BAT
Best available techniques/technology to minimize pollution, can be implemented effectively and are economically and technically viable. Sometimes known as BATNEEC – best available techniques not entailing excessive cost.

BPEO
Best practicable environmental option. The outcome of a systematic consultative and decision making procedure which emphasizes the protection of the environment across the media of land, air and water. It goes beyond the BEO concept and includes an economic analysis. BPEO procedures provide the most benefit, or least damage, to the environment as a whole, at acceptable cost, in the long term, medium term as well as the short term.

Bruntland Report
Report of the 1987 World Commission on Environment and Development which established the concept of 'Sustainable Development' ie *'Development that enables present generations to meet their needs without impairing the ability of future generations to meet their needs'.*

Certification
Procedure by which a third party provides written assurance that a product, process, service or management system conforms to specified requirements.

Certified
The EMS of a company, location or plant is certified for conformance with ISO 14001 after it has demonstrated such conformance through an audit process.

Client
Organization commissioning the audit (ISO 14010).

Compliance
A positive indication or judgement that *the supplier* of a product or service has met the requirements of the relevant specifications.

Conformance

A positive indication or judgement that a *product* or *service* has met the requirements of the relevant specifications, contract or regulation; also the state of meeting the requirements.

Contaminated land

Land which has absorbed a substance or substances that could pose a hazard to health and the environment.

Continual improvement

Process of enhancing the EMS to achieve improvements in overall environmental perform-ance, in line with the organization's stated environmental policy (ISO 14001).

Contractor

Any external organization that provides a product, good or service to a customer in a contractual situation.

Corrective action

An action taken to eliminate the causes of an existing non-conformity, defect or other undesir-able situation in order to prevent its recurrence.

Cradle to grave

See Life cycle assessment below.

Critical load

A quantitative estimate of exposure to pollutants below which no significant harmful effects are believed to result.

Customer

Ultimate consumer, user, client, beneficiary or second party.

Directive

A European Union legal instrument identifying an outcome binding on all Member States yet leaving the method of implementation to national governments through national legislation.

Discharge consent

A licence granted by a regulatory body to permit the discharge of effluent of specified quality and volume.

Eco-labelling

A logo which confirms a product meets environmental criteria as set out under European Union Regulation relating to a product's entire life cycle.

Effluent

Liquid waste released to the environment from industrial, agricultural or sewage treatment plant.

EMAS

The European Eco-Management and Audit Scheme is a voluntary European scheme that will evaluate a participant's system for good environmental management and publish the results.

EMS audit

A systematic and documented verification process to obtain and evaluate evidence objectively to determine whether an organization's EMS conforms to the EMS audit criteria set by the organization, and to communicate the results of this process to management (ISO 14001).

EMS audit criteria

Policies, practices, procedures or requirements, such as those covered by ISO 14001 or EMAS and, if applicable, any additional EMS requirements against which the auditor compares collected evidence about the organization's EMS (ISO 14011).

Environment

The surroundings in which an organization operates, including air, water, land, natural resources, flora, fauna, humans and their interrelation. Note: Surroundings in this context extends from within the organization itself to the wider focus of the global system.

Environmental aspect

Element of an organization's activities, products and services that can interact with the environment (ISO 14001).

Environmental audit

Systematic, documented verification process of objectively obtaining and evaluating audit evidence to determine whether specified environmental activities, events, conditions, management systems or information about these matters conform with audit criteria, and communicating the results of this process to the client (ISO 14010).

Environmental impact

Any change to the environment, whether adverse or beneficial, wholly or partially resulting from an organization's activities, products or services (ISO 14001).

Environmental liability

Liability for environmental damage which may arise from either statute or law.

Environmental management system (EMS)

Organizational structure, responsibilities, practices, procedures, processes and resources for developing, implementing, achieving, reviewing and maintaining the enviromental policy (ISO 14001).

Environmental objective

Overall environmental goal, arising from the environmental policy, which an organization sets itself to achieve, and which is quantified where practicable by setting realistic targets (ISO 14001).

Environmental performance

The measurable results of the EMS, related to an organization's control of its environmental aspects, based on its environmental policy, objectives and targets (ISO 14001).

Environmental policy

Statement by the organization of its intentions and principles in relation to its overall environmental performance, which provides a framework for action and for the setting of its environmental objectives and targets (ISO 14001).

Environmental target

Detailed performance requirement, quantified wherever practicable, applicable to the organization or parts thereof, which arises from the environmental objectives and that needs to be set and met in order to achieve those objectives (ISO 14001).

Follow-up audit

An audit whose purpose and scope are limited to verify that corrective or preventive action has been accomplished as scheduled and, where appropriate, to determining that the action effectively prevented recurrence.

Global warming

Climatic change resulting from changes in atmospheric conditions caused by increased levels of pollutants.

Greenhouse effect

The process by which gases in the atmosphere, such as carbon dioxide, methane and CFCs, allow solar radiation to reach the earth but absorb heat radiation emitted from the earth. The net effect is an increase in the average temperature of the earth.

Ground water

Water held in the ground in aquifers.

Heavy metals

A collective term used for metals with the potential to cause harm when they are released into the environment. Typically it includes mercury, lead and cadmium, as well as zinc, chromium and certain other metals with wide industrial use and potential toxic effects.

Hydrocarbons

Chemical compounds composed only of carbon and hydrogen. Many can react with nitrogen dioxide under the influence of sunlight to produce photochemical oxidants.

ISO 14001

ISO 14001 – *Environmental Management Systems, Specifications with guidance for use* – is the international equivalent of BS 7750 which it superseded in September 1997.

Interested party

See Stakeholder.

Lead auditor (environmental)

Person qualified to manage and perform environmental audits (ISO 14012).

Life cycle assessment (LCA)

A process for evaluating the environmental burdens associated with a product, process or activity by identifying and quantifying the energy and material use and environmental releases. The assessment includes the entire life cycle of the product, process or activity and encompasses extracting and processing the raw materials, manufacturing, transportation, distribution, reuse, maintenance and ultimate disposal. Also known as 'resource and environmental profile analysis (REPA)' (qv), 'eco-profile analysis', 'eco-balancing' or 'cradle to grave' analysis (ISO 14040).

Non-conformity/Non-conformance

The non-fulfilment of a specified requirement (ISO 8402, Clause 3.20).

Ozone depleting chemicals

Chemicals, such as chlorofluorocarbons (CFCs) and halons, which have the potential to deplete the ozone concentration in the stratosphere. While these chemicals are inert (chemically unreactive) in the troposphere, they can be broken down in the stratosphere to release chlorine and bromine which take part in the destruction of ozone. This ozone provides a protective barrier against harmful ultraviolet radiation from the sun.

Ozone hole

The common term for the depletion in ozone in the stratosphere over the Antarctic region.

Particulate matter

Generally used for solid particles (dust) emitted from processes and dispersed in the atmosphere. The term can also include liquid droplets.

Polluter pays principle

The principle that the cost of preventing pollution or of minimizing pollution damage should be met by those deemed responsible for the pollution.

Precautionary principle

Where significant enviromnental damage may occur, but scientific knowledge is incomplete, decisions made and action taken should err on the side of caution.

Prevention of pollution

Use of processes, practices, materials or products that avoid, reduce, or control pollution, which may include recycling, treatment, process changes, control mechanisms, efficient use of resources and materials substitutions (ISO 14001).

Raw material

Primary or secondary recovered or recycled material used in a manufacturing system to produce a product (ISO 14040).

Red list

Twenty-three toxic substances, defined in the UK as those presenting the greatest potential threat to the aquatic environment. The list includes several heavy metals (qv), certain pesticides and solvents.

Registration
Procedure by which a body indicates relevant characteristics of a product, process or service, or particulars of a body or person, and then includes or registers the product, process, or service in an appropriate publicly available list (ISO/IEC Guide 2).

Regulation
In the European context this can mean European Union legislation having direct legal force in all Member States.

Requisitory assessment level (RAL)
The long-term average concentration of a substance which, for the purpose of assessing the BPEO (qv), the regulatory body regards as the maximum value permissible in the environmental medium concerned at that location.

Responsible Care
An initiative which began in Canada in the mid 1980s. It provides comprehensive guidelines for EMSs adopted by the Chemical Manufacturers Association (CMA) in 1988. Participation by individual businesses is an obligation of membership in the CMA.

Risk assessment
A method whereby the risks associated with a particular site are identified and quantified.

Root cause
A fundamental deficiency that results in a non-conformance and must be corrected to prevent recurrence of the same or similar non-conformance.

Specification
The document that prescribes the requirements with which the product, service or management system must conform.

Stakeholders
Those groups and organizations having an interest or stake in an organization's EMS programme (for example: regulators, shareholders, customers, suppliers, special interest groups, residents, competitors, investors, banks, media, lawyers, insurance companies, industry groups, trade unions). Sometimes referred to as 'interested parties'.

Stratosphere
The level of the atmosphere above the troposphere and about 15–50 km above the earth's surface.

Subcontractor
An organization that provides a product or service to a client organization.

Supplier
An organization that provides a product to the customer.

Sustainable development
Development that meets the needs of the present and allows economic growth without compromising the ability of future generations – through depletion of resources including energy – to meet their own needs.

System boundary
Interface between the product system being studied and its environment or other systems (ISO 14040).

Third party
Person or legal entity recognized as being independent of the parties involved in the sale of a product, goods or service. Suppliers or producers are the first party and consumers the second party (ISO 14024).

Verification
Process of authenticating evidence. The act of reviewing, inspecting, testing, checking, auditing or otherwise establishing and documenting whether items, processes, services or documents conform to specified requirements.

Volatile organic compound (VOC)
An organic compound other than methane that is capable of producing photochemical oxidants.

Waste
Any output from the product system that is disposed of.